FOOD ALLERGY AND INTOLERANCE

Current Issues and Concerns

Edited by
Victoria Emerton

This edition first published 2002 by
Leatherhead Publishing
a division of
Leatherhead International Ltd
Randalls Road, Leatherhead, Surrey KT22 7RY, UK
URL: http://www.lfra.co.uk

This edition published 2002 by Royal Society of Chemistry,
Thomas Graham House, Science Park, Milton Road, Cambridge CB4 0WF, UK
URL: http://www.rsc.org
Registered Charity No. 207890

Special Publication No. 285

ISBN: 0 85404 881 2

A catalogue record of this book is available from the British Library

Typeset by Leatherhead International Ltd.
Printed in the UK by IBT Global, 1B Barking Business Centre, 25 Thames Road, Barking, London IG11 0JP
and bound by MPG Books Ltd, Bodmin, Cornwall, UK

CONTENTS

CONTRIBUTORS

Professor John Warner has been Professor of Child Health in Southampton since 1990. He trained in Great Ormond Street before becoming a consultant at the National Heart and Lung Institute and the Brompton Hospital in 1979. He has published widely in the area of allergy and asthma (over 200 papers). Recent emphasis has been on the foetal and infant origins of allergic disease and he is currently Editor-in-Chief of the Journal of Pediatric Allergy and Immunology.

Dr Rebecca Dearman joined the Central Toxicology Laboratory in 1988. Her current position is Head of Immunology Research and her active research interests include all aspects of allergy, particularly the cellular and molecular regulation of food allergy, chemical respiratory sensitisation and contact allergy, cytokine biology, dendritic cell biology and immunotoxicology. Dr Dearman has published over 120 research papers, review articles and book chapters. She is an active member of the British Society for Immunology, the British Toxicology Society and the ILSI Protein Allergy Subcommittee.

Dr Jean-Michel Wal is currently Director of Research at INRA, and head of the laboratory of food allergy, a joint venture between INRA and CEA. Dr Wal is currently participating in many current EU-funded research programmes, and is Chairman of the Working Group on Safety Testing of Transgenic Foods, as well as being a member of several official scientific and advisory committees, including the French National Council for Food, and EU Scientific Committee for Food.

Having moved back to Southampton from Great Ormond Street Hospital in January of 2001, **Dr Jonathan Hourihane** is Assistant Director, Wellcome Trust Clinical Research Facility, where he is establishing an independent research programme, looking at cellular features of the immune response to food allergens, using carefully characterised subjects. The standardisation of diagnostic and research food challenges is a particular focus of Dr Hourihane's clinical research.

Dr Stephan Vieths is the Research Coordinator for the Paul Ehrlich Institute, which he joined in 1995, and Associate Professor of Food Chemistry at the Johann Wolfgang Goethe University in Frankfurt. Dr Vieths' work currently focuses on the production and characterisation of recombinant food allergens, and on the detection of allergen traces in processed foods. Dr Vieths has published 50 original articles on food allergy.

Dr Cristina Pascual joined the Hospital Infantil in 1980, where she is currently a paediatric allergologist. She was the coordinator of the Food Allergy Committee of the Spanish Society of Allergy and Clinical Immunology from 1992-1996, and has been a member of the Committee of Adverse Reactions to Foods of the American Academy of Allergy, Asthma and Immunology since 1998, and recently became a member of the executive board of the European Academy of Allergology and Clinical Immunology.

Dr Moneret-Vautrin has been Head of the Department of Internal Medicine, Clinical Immunology and Allergology at Hopital Central in Nancy since 1990. She has published widely in the field of food allergy and is referee for a number of key allergy titles, including Clinical and Experimental Allergy and Allergy journals.

Mrs Hazel Gowland has been allergic to nuts since suffering her first reaction in 1960, and has survived a number of life-threatening reactions since. She has worked with the Anaphylaxis Campaign since its earliest days in 1994. Mrs Gowland is now committed to an increasingly busy programme of factory and foodservice site visits to exchange ideas for risk minimisation. She regularly speaks at conferences, exhibitions and training events and writes articles and information leaflets. In April 2000, Hazel was a finalist in the first BBC Radio 4 Food Programme Awards in the Campaigner/Educator category.

Dr Tara Dean began her research career in the Paediatric Allergy and Immunology Unit at the Royal Brompton and National Heart Hospital, London. Dr Dean joined the University of Southampton as a Senior Research Fellow in the Child Health Department in 1992. She is honorary Deputy Director at the Asthma and Allergy Research Centre on the Isle of Wight, and in her current post as Reader in Health Services Research at the University of Portsmouth, she provides support to all healthcare professional undertaking research and development in West Sussex, the Isle of Wight, Portsmouth and south east Hampshire.

Professor Peter Howdle is a Consultant Gastroenterologist at St James's University Hospital, Leeds. He has a long-standing clinical and research interest in coeliac disease. He has published widely on the immune aspects of the disease, and on in vitro models. He edited "Coeliac Disease" for Balliere's Clinical Gastroenterology, in June 1995. He is one of the Medical Advisers to the Coeliac Society of the UK.

After graduating as a specialist in hepato-gastroenterology in 1986, **Dr Philippe Marteau** obtained a PhD in microbiology in 1994, and has been a Professor of Gastroenterology since 1997. Dr Marteau now works as a clinical gastroenterologist in the gastroenterology department of the Hôpital Européen Georges Pompidou, Assistance Publique des Hôpitaux de Paris, and taught gastroenterology and intestinal physiology at the Paris V University. His main research interests are intestinal ecology, probiotics, lactose and non-digestible carbohydrates and inflammatory bowel disease. He has written more than 100 papers in peer-reviewed journals, and approximately 20 book chapters.

Mr Robert Palgrave was diagnosed with coeliac disease in 1998, and started work with the Coeliac Society in March 2001 as Editor on the publication "Crossed Grain" for Members. Crossed Grain currently has a readership of approximately 50,000.

Fiona Angus has worked for Leatherhead since 1987. She has undertaken a variety of roles during this time but, since 1993, has been closely involved with the issue of food allergy and intolerance. Over the past 8 years, she has provided consultancy to companies, managed a food intolerance databank, and coordinated a number of MAFF and EU projects related to food allergy and intolerance.

Professor Paul Davis is a senior scientist in Unilever Research, where he has worked for many years on aspects of applied immunology. He is Visiting Professor of Applied Immunology at the University of Kent at Canterbury. His current interests include mechanisms of biomolecular recognition, biosensors for volatile molecules, the chemistry of antigen modification and the limits of immune recognition.

Dr Clare Mills graduated from the University of Bristol and has a PhD at the University of Kent. After a year working at the Department of Health, she joined the Institute of Food Research working in the Biorecognition and Immunology Group. In 1997, she moved to work with Professor Belton and in 1999 became Head of the Physical Biochemistry Group. She has a long-standing interest in the role of food protein properties in food functionality, particularly in cereal- and legume-based foods, together with the way in which plant protein structure component interactions affect food protein allergenicity.

Dr Sabine Baumgartner obtained her PhD in the structural parameters of polysaccharides in aqueous solutions from the University of Agriculture in Vienna and has worked in the Center for Analytical Chemistry at the Institute of Agrobiotechnology in Tulln. She is currently head of the Working Group for Biochemical Analysis.

Jenny Roberts has been Company Dietitian for Iceland Foods Plc for just over a year. In this time a major part of her role has been to drive allergen and nutrition supplier policies, as well as advances in product labelling and the formulation of the first 'nut list' for customers.

Professor John Warner
Professor of Child Health
Allergy and Inflammation Sciences
University of Southampton
Southampton
SO16 6YD

Dr Rebecca J. Dearman, Professor Ian Kimber
Syngenta Central Toxicology Laboratory
Alderly Park
Macclesfield
Cheshire
SK10 4TJ

Dr Jean Michel Wal
Laboratoire d'Immuno-Allergie Alimentaire
INRA, CEA de Saclay
91191 Gif-sur-Yvette
France

Dr Jonathan O'B Hourihane
Senior Lecturer, Infection, Inflammation and Repair Division,
Assistant Director,
WellcomeTrust Clinical Research Facility
Mailpoint 218
Southampton University Hospitals NHS Trust
Tremona Road, Southampton
SO16 6YD

Professor Dr Stefan Vieths, Dr Jörg Kleine-Tebbe
Department of Allergology
Paul Ehrlich Institut
Paul-Ehrlich-Straße 51-59
D- 63225 Langen
Germany

Dr Cristina Pascual, Dr M.T. Belever, Dr A. Valls, Dr M. Martin-Esteban
Hospital Infantil La Paz
Paseo de la Castellana, 261,
28046 Madrid
Spain

Dr Denise-Anne Moneret-Vautrin
Department of Clinical Immunology and Allergology
Hopital Central 29
Avenue de Lattre de Tassigny
54035 Nancy
France

Mrs Hazel Gowland
Allergy Action
23 Charmouth Road
St Albans
Herts
AL1 4RS

Dr Taraneh Dean
Portsmouth Institute of Medicine, Health and Social Care
University of Portsmouth
Portsmouth
Hampshire

Dr Phillippe Marteau
Service d'Hepato-Gastroenterologie
Hopital Europeen Georges Pompidou
20 Rue de Leblanc
75908 Paris Cedex 15
France

Professor Peter D. Howdle
Academic Unit of Medicine
St. James's University Hospital
Leeds
LS9 7TS

Mr Robert Palgrave
Coeliac UK
PO Box 220
High Wycombe
Bucks
HP11 2HY

Mrs Fiona Angus
Leatherhead Food Research Association
Randalls Road
Leatherhead
Surrey
KT22 7RY

Ms Jenny Roberts
Iceland Foods plc
Second Avenue
Deeside Industrial Park
Deeside
Flintshire
Wales
CH5 2NW

Professor Paul Davis, Dr C.M. Smales, Dr. D.C. James
Insense Ltd.
Colworth House
Sharnbrook
Bedford
MK44 1LQ

Dr Clare Mills
Institute of Food Research
Norwich Research Park
Colney
Norwich
NR4 7UA

Dr Sabine Baumgartner
IFA-Tulln, Center for Analytical Chemistry
Konrad Lorenz Straße 20
A-3430 Tulln
Austria

PREFACE

The prevalence of allergy appears to be increasing, with some interesting data indicating that trends in allergy to different foods are influenced by common foods in the diet of particular countries. Consumers are more aware of allergy than ever, with retailers launching special ranges of foods suitable for people with particular allergies, and allergy featuring regularly in the media.

This book contains the lectures that were given at the conference "Food Allergy and Intolerance 2001 – Current Issues and Concerns", held in London at the end of 2001. The conference gathered together many leading experts in food allergy from around the world, and covered latest developments on all the major allergens such as nuts, milk, soya, seafood, and sesame, also including foods to which intolerance is common - for example, gluten, lactose and food additives. The reasons for the increasing incidence of allergy are still not fully understood, or the mechanisms of sensitisation to particular foods, either in the womb or in adulthood; research into the mechanisms and factors affecting allergen sensitisation was presented, and the impact of food allergy and intolerance on the consumer and the consumer's family was clearly outlined by consumers allergic to nuts and intolerant to gluten.

The advent of genetic modification and development of novel foods opens up the potential for introduction of new allergens into the food chain, or the transference of allergenic components from one food type to another. The characteristics of proteins that cause an allergic response were discussed, together with the steps currently in place to assess the allergenicity of novel foods. The latest research into the changes in protein structure brought on by processing, and the effects on allergenicity of proteins were discussed, as well as the development of dipstick assays for detection of allergens.

Food allergy has had a significant impact on the food industry with respect to food labelling, and the importance of Good Manufacturing Practice to minimise or eliminate cross contamination with allergens. The issues and practical problems that food allergy causes the food industry were outlined, and steps being taken by one food retailer to help its food-allergic or -intolerant customers and to provide information on food allergy and intolerance were presented.

This book is not only intended as a reference for those who were present at the conference, it is also an informative resource for all who are interested in food allergy and intolerance, and want to keep up to date with latest developments in this field.

My sincere thanks go to all the authors.

Victoria Emerton

1. THE EARLY LIFE ORIGINS OF FOOD ALLERGY

Professor John Warner

1.1   Introduction

Food allergy is certainly a major concern for many, and it captivates the media. Of course, it is important to distinguish what is fact from what is fiction and many parents believe their children to have a problem with foods when in fact it is just a behaviour problem or something that is totally independent of the immune system. This paper will concentrate on immunologically mediated reactions to food that occur usually very rapidly after exposure and are predominantly associated with the generation of an allergy antibody that is an IgE antibody to the food.

Allergic diseases associated with the generation of IgE antibody, whether they be associated with inhalants or with foods, have been increasing in prevalence very considerably over the last two or three decades in affluent communities, and now similar increases are occurring in the developing countries as Western dietary and lifestyle habits are adopted.

As a consequence, there is increasing morbidity from all of these illnesses and a major socio-economic cost, which clearly has to be of considerable concern to the health services. This has resulted in exponential increases in the need for various drugs to treat the condition and there are no treatments available at present to cure allergy. Time sometimes modifies the severity of the allergic response, but there are no medical cures for food allergy or any other allergic diseases at present.

In order to understand the immunology of food allergy, and indeed all allergic diseases, one must understand the concept of the balance between T helper 1 (Th-1) and T helper 2 (Th-2) cells; that is, there are some cells, known as lymphocytes, that are the controllers, the conductors, of the immune system, and, by virtue of the chemical messengers that they produce, orchestrate the immune response. Th-1 lymphocytes generate a range of chemical messengers, or cytokines, of which the most important are interferon-gamma and interleukin 2 (IL2), which orchestrate a normal immune response to bugs, to micro bioorganisms. They initiate an immune response that gets rid of bacterial and viral infections. They will also stimulate elimination of early cancerous-type cells.

The Th-1 immune response is also sometimes associated with disease, and it is up-regulated, or increased, in auto-immune diseases, such as rheumatoid arthritis, and also is the key response that results in transplant rejection, but can be controlled by giving immuno-suppressive drugs. It is also involved in early miscarriages in initiating rejection of the foetus by the maternal immune system.

The Th-2 response is associated with allergy. Th-2 levels are elevated in allergic disease and the Th-2 cells generate cytokines, of which IL-4, 5, and 13 are particularly important. Th-2 cells are also associated with beneficial responses to eliminate parasites from the gut and are potentially significant in preventing foetal rejection and sustaining pregnancy. Understanding this mechanism is important for us in order to be able to establish the processes in early life that result in allergic disease.

1.2   Production of IgE Antibodies

The Th-2 cytokines, IL-4 and IL-13, promote the production of the allergy antibody IgE, whereas interferon gamma prevents the production of IgE. There are many other cytokines involved in a network, controlling each other, and IL-10 happens to be one that prevents the production of interferon gamma. The balances of the various T helper cells and cytokines are important, and in allergic disease the balance is towards Th-2.

1.3   *In utero* Allergen Sensitisation

Does allergic disease start in pregnancy? In some respects it does, because pregnancy itself is an allergic state. Because the foetus represents foeto-paternal as well as foeto-maternal antigens, the maternal immune system should, under normal circumstances, reject the foetus. Indeed, this does occur under some circumstances and that results in early and sometimes recurrent miscarriage.

The placenta acts in part as a barrier between the mother and the foetus and also as an orchestrator of the mother's immune responses. There are many cells in the placenta generating IL-13, and the amnion cell produce IL-4.

Amniotic fluid samples contain a lot of IL-10, which prevents the production of interferon-gamma. Prostaglandin E2 is very much in evidence in the amniotic fluid, and that promotes the production of IL-4 and other Th-2 cytokines, whereas those associated with Th-1 activity are usually not present at all, such as interferon-gamma and IL-12. So there is an imbalance here, with Th-2 activity being up-regulated. The effect of that is to dampen the mother's Th-1 response, which is a normal immunising tissue-rejecting, foetal-rejecting cell-mediated immune response. This is a mechanism that protects the foetus, by the decidual tissues producing these cytokines and dampening the maternal immune response.

There are many other complexities to this interaction and this is but one component of the mechanism that protects the foetus. Clearly, from the point of view of the allergist this is the mechanism that is of interest. The question is, if this mechanism is having a dampening effect on the mother, what is it doing to the foetus? Is it making the foetus begin to move towards an allergic response? There is some evidence that this is the case (1). Furthermore, the foetus is also exposed to allergens to which the mother is also exposed. House dust mite and food allergens can be detected in the amniotic fluid and the foetal circulation. Thus the foetus will be swallowing allergen and cytokines that promote an allergic response (2).

The first set of cells which amniotic fluid factors will encounter in the foetal gut, pick up foreign proteins, antigens, or, in the case of the allergist, allergens. These cells are known as antigen presenting cells, which engulf protein particles, process them, and then present them to the immune system to induce a sensitivity. On examining the gastrointestinal tract of the foetus at just 11 weeks' gestation (at the end of the first trimester of pregnancy), the foetal gut is loaded with antigen presenting cells. So, throughout the foetal gut, there are numerous cells that are capable of picking up foreign proteins and processing them to present to the immune system. These cells will present peptide fragments to the lymphocytes, and orchestrate a switch-on, a sensitisation of the lymphocytes, and then the lymphocytes will pass on the message to B lymphocytes; these B cells, once they have received the message from T cells, will begin to generate antibodies, and obviously those of greatest concern with respect to allergy are IgE antibodies.

By looking at the maturity of the antigen presenting cells in the foetal gut, it can be demonstrated that, by 16 to 17 weeks' gestation, there are cells mature enough to pick up antigen, and indeed there is evidence that they have done so, presented them to T cells, which have been sensitised, and in turn switched on B cells to produce antibodies. All of that can happen from the middle of the second trimester of pregnancy. So the foetus is not as immunologically naïve as one might think (3). Because of the environment in which the foetus is swimming, the immune response will be biased towards an allergic type of response.

What the foetus does to try to balance up that response, and not be totally committed to an allergic response is to generate interferon -gamma spontaneously. The foetus' circulating cells produce quite large quantities of interferon -gamma, which is released spontaneously into the circulation. However, both genetic and intrauterine environmental factors will compromise the foetus' capacity to generate interferon gamma and balance the response.

In the third trimester of pregnancy however, there is another route of exposure to allergen, and that is by active transport complexed with immunoglobulin G (IgG). This is a protective antibody that passes from the mother to the baby to build up its passive immunity for after birth. IgG is sometimes complexed with allergens, and IgG ovalbumin, β-lactoglobulin, cat and house mite complexes with IgG are known to cross the placenta (4).

The implications of this in relation to disease development needs to be understood. It is quite clear, from blood samples taken from foetuses at different periods of gestation, that by about 22 weeks' gestation, some foetuses are beginning to respond to allergen, and this is reflected by the lymphocyte's ability to proliferate when stimulated with allergens such as β-lactoglobulin from cows' milk. If there is a ratio of stimulated over unstimulated cells of higher than two, that is a significant response. By somewhere between 22 and 24 weeks, from virtually any allergen or antigen to which the mother and, therefore, the foetus have been exposed, it is possible to observe responses developing that increase progressively through pregnancy. By birth the overwhelming majority of newborn babies have an immunological-specific response to a host of antigens to which the mother has been exposed.

However, there is evidence that antigens and allergens that cross the placenta complexed with IgG have an effect of down-regulating, (i.e. reducing) the immune response, and that these complexes switch off the sensitisation or prevent it from occurring in the first place. There is an inverse correlation between the IgG antibody levels at birth, which have come from the mother, and how those levels relate to subsequent sensitisation (5).

We have observed levels of IgG-ovalbumin antibodies in the cord blood that came from the mother, and correlated this with whether the baby, in the first 18 months of life, developed a positive skin prick test to egg. What resulted was that, if the IgG antibodies were very low, which probably implies that the mother had had very little exposure to ovalbumin, there were no children who become sensitised subsequently to egg. But, oddly, at the very high levels of exposure, where there was a significant amount of IgG antibody, which picked up all the complexed ovalbumin that the mother ate, and therefore exposed the foetus only to complexed allergen, again the majority of children did not become sensitised to egg, whereas, in the middle range, where there was a small amount of IgG present, but probably not enough to mop up all the ovalbumin that got into the mother's circulation if she ate any, the major sensitisations occurred. Combining the information, we believe that amniotic fluid allergen exposure in low doses sensitises the foetus, while high doses complexed with IgG in late pregnancy switch off sensitisation.

1.4   Allergen Avoidance

What happens if we reduce the mother's intake of allergen through pregnancy? One could suggest that there is a risk of shifting the population into the middle range, where sensitivity occurs. Allergen avoidance is not necessarily the answer to preventing allergy, because there are issues about high-zone tolerance versus low-dose no-sensitisation that are very complex. The effectiveness of antenatal allergen avoidance with respect to the dose and timing of exposure is still unclear. More research is needed, and, for the moment, no recommendations are being made unless total avoidance is possible.

It can be debated whether total avoidance is possible for foods such as peanuts and tree nuts, compared with egg and milk; low-dose exposure may be more likely to sensitise, particularly if it occurs in the second trimester of a pregnancy, compared with high-dose exposure. In other words, there is no firm evidence of the action that should be taken in relation to avoidance of specific allergens in pregnancy.

1.5   Genetic Influences

Most newborns are sensitised to food allergens; what tips them over into being much more committed to an allergic response and getting allergic disease? One factor is obviously genetics. It is remarkable that if the baby inherits allergy-linked genes from the mother, then it is four to five times more likely to manifest food allergy and allergic eczema by one year of age than if it inherits allergy-linked genes from the father (6). One explanation could be that this could be some form of genetic imprinting, or, far more likely, it is something to do with the environment that is created by the allergic mother for her allergic-prone foetus that makes her more likely to produce allergic off-spring.

It has been observed that the cytokine in amniotic fluid that suppresses the anti-allergic, interferon gamma production, is very much higher in allergic mothers than in non-allergic mothers. So the amniotic fluid has more allergy-promoting cytokines if the mother is allergic, compared with the father. This is one explanation for this change in environment created by the allergic mother.

The allergic mother will inevitably have higher levels of IgE herself. There is a direct relationship between the mother's IgE level in her circulation and the level of IgE that can be found in the amniotic fluids. IgE diffuses passively across the amnion and small quantities of IgE can be detected in amniotic fluid.

If IgE is present in the amniotic fluid, together with allergen and Th-2 cytokines, and the foetus swallows those, the IgE will stick to receptors on the antigen presenting cells and facilitate an immune response, by "antigen focusing". IgE will help the immune system to respond to 100- to 1,000-fold lower concentrations of allergen. So the presence of the IgE getting across to the foetus from the mother will increase the chance of that foetus developing an immune response at an early stage in pregnancy. This antigen focusing is thought to be the main mechanism by which an allergic mother primes her foetus to be more likely to respond subsequently in an allergic way (7).

There are other factors involved, and research has turned to a molecule called CD14, which helps the immune system to respond to bacteria, producing a normal immunising response - in other words, a Th-1 response, generating interferon gamma, which prevents allergy. It has been found that the level of CD14 in the amniotic fluid is lower when babies have subsequently become allergic and developed eczema within the first 12 months of life, compared with babies who have not subsequently developed eczema. CD14 probably comes from the mother into the amniotic fluid, is swallowed by the foetus, and combines with various bacterial antigens, which will switch an immune response to an appropriate immunising response (8).

It is not surprising that if T lymphocytes from a cord blood sample from a newborn baby proliferate avidly in response to a food antigen, such as β-lactoglobulin, and at the same time there is an imbalance of the chemical messengers, and low interferon gamma production, then that neonate is far more likely to go on to develop milk allergy and atopic eczema. A direct relationship has been demonstrated in that, at birth, not only can an allergy-prone individual be identified, but also to which allergens he/she is likely to be sensitive (9).

1.6   Postnatal Allergy Development

How can the immune system be taught to respond to allergens more appropriately after the child is born? A whole series of factors suggest that hygiene, or factors that reduce exposure to microbial organisms early on, confer a significantly increased risk of developing allergy. The so-called hygiene hypothesis is widely recognised; numerous studies have shown from very different perspectives that living in a clean and ultra-hygienic environment, with reduced exposure to ordinary bugs, is associated with a higher incidence of allergy (10). Birth order is also a significant factor in incidence of allergy: first-born children have a higher incidence of allergy than second- and third-born children, because the second- and third-born succumb earlier to infections that are delivered to the home by the first-born. Geographical variations suggest that, in environments where children are exposed to more bugs, they have a lower incidence of allergy. In Japan, and a number of other countries, there have been studies showing that, if people have a good response to tuberculo-protein, in other words if there is a higher local prevalence of tuberculosis, then they have less allergy, and there is an inverse relationship between the response to tuberculin and prevalence of allergy. In situations where measles or hepatitis A occurs, and other organisms associated with lower standards of hygiene are present, there is less allergy; for example, children born on farms suffer less allergy. In areas where the early and frequent use of antibiotics is common, the incidence of allergy is higher. Use of oral antibiotics in the first 2 years of life confers a considerably increased risk of subsequently being allergic, and this risk is as great as the increase in risk of allergy associated with having a mother who is allergic. Recent studies have focused on gut flora, and have shown that various bugs in the gut are different in those who are allergic compared with those who are not allergic.

In conclusion, genes and the *in utero* environment are significant, but postnatal early exposure to microbial organisms is clearly important in teaching the immune system to respond normally.

In pregnancy, the foetus is in a germ-free environment and has low-concentration allergen exposure, particularly in the second trimester of pregnancy, with Th-2 allergy promoting mediators in the amniotic fluid switching on an allergic immune response. Postnatally, though, this rapidly switches off in the non-allergic individual, probably because of exposure to a high allergen load in a germ-laden environment, whereas the allergic individual exhibits a progressively reinforced allergic response. For some, transient food allergy switches off after the first year, and the circumstances that cause the switching off of immune response and what is involved with perpetuation of allergy require further study.

It is clear that the important postnatal events regarding allergy are related to what microbiological organisms the individual is exposed to, and when, and what the allergen load is, against a background of genetic predisposition.

1.7   Factors Influencing Allergy

1.7.1   Early immunisation

It is possible that, from the point of view of therapeutic intervention, the answer may well lie in the soil. There is a suggestion that early BCG vaccination against tuberculosis can protect against allergy. A study in Guinea-Bissau looking at early BCG immunisation versus no BCG or later BCG immunisation showed a halving of the prevalence of allergy over the first years of life if BCG was administered early (11).

BCG does not come from the soil, but there is another mycobacterial organism, *Mycobacterium vacci*, that does, and that organism is now being used in immunising trials as a way of trying to switch off allergy by orchestrating this switch away from Th-2 and towards Th-1, and the early studies look very promising.

1.7.2   Gut microflora

The other therapeutic area with potential is that of gut organisms. The hypothesis that having the right microbial flora in the gut might change allergic status is of intense interest. There is now one study published that has looked at giving *Lactobacillus* GG as a probiotic to newborn babies, up to 6 months of age. The children, who all came from high-risk families, where there was a lot of atopy in the family already, were then monitored, and results showed that, in the group given probiotic, the subsequent rate of atopic eczema was significantly reduced up to 2 years of age, compared with the group that was given the placebo (12). This result goes some way to supporting the Th-1 and Th-2 hypothesis and the hygiene theory. The only anomaly with this study, is that there was actually no difference in IgE or positive skin prick tests in these babies. Dosing with probiotic prevented disease, but not allergy, which suggests that the complexities of the interaction between Th-1 and Th-2 are far greater than previously thought. It could be that another set of regulatory cells which control both Th-1 and Th-2 activity may in the end be important as well.

1.7.3   Foetal growth rate

An unusual relationship has been found between foetal growth and the potential for becoming allergic. The link between head circumference at birth and a raised IgE antibody subsequently has been shown. This relationship is in adults, and it has been shown that a high percentage of adults had high IgE levels if they had big heads at birth. The hypothesis behind this relationship is that, in an affluent society, where allergy is increasing, and where there is excellent nutrition, the foetus is programmed to grow very rapidly during the early stages of pregnancy because of the good nutrient delivery from the mother. Once that programming has occurred, the foetus will continue to grow at that rate. Of course, later in pregnancy, it will have a high nutrient demand, which, paradoxically, in the affluent communities, will not be met. In a poor community, with poor nutrition, the foetus is programmed to grow slowly and will, therefore, have a lower nutrient demand and will sustain its growth at the rate set early in pregnancy. Here we have a situation where the foetus is suddenly in a position where the nutrient delivery is not good enough to sustain growth. So what happens is that the brain and head continue to grow at the expense of the body, which means that the baby has a big head and normal birth weight, but, of course, the poor nutrient delivery to other tissues in the body can produce a modification of the immune responses. It is becoming apparent that Th-1 immunising non-allergy responses are more susceptible to being switched off in adverse circumstances than Th-2. So there is a selective reduction in Th-1 responses and, therefore, allergy is associated with having a large head circumference, not because large head circumference causes allergy, but because of an indirect relationship. This may explain the increase in allergy in affluent societies (10). If this is the case, the next step is to begin to understand what nutrients may be important in controlling the immune system - for example, are antioxidants, or fish oils, which are incorporated into cell membranes, and are important in relation to signal transduction important? Are they particularly important in relation to nutrient delivery to the foetus for sustaining foetal growth and immune development? This is still undetermined and there are some interesting trials going on at the moment that are looking at various forms of supplementation during pregnancy, particularly with antioxidants and fish oils to see whether they have any impact on allergy.

1.8   The Allergic March

Ultimately, what is key is that food allergy is but the start of the whole sequence of events. The food-allergic individual quite commonly has eczema, and quite commonly goes on to develop asthma and allergic rhinitis. That is the beginning of the allergic march. So, in understanding food allergy, we should, in the end, understand more about allergy in general. Being allergic to egg, for instance, in the first years of life, confers a very considerable increased risk of developing asthma, compared with not being allergic to egg. An example of this is provided by a study in which babies all had eczema; some were allergic to egg and some were not. The children were followed up over 3 years, and 55% of the egg-allergic eczematous children had asthma 3 years later; 40% of those who were not egg-allergic went on to develop asthma. Interestingly, that was not the case for milk, but, if the infants were allergic to inhalants, such as grass pollen, dust mite, or cat dander, then they had a very much higher risk of developing asthma. Furthermore, amongst those children who were allergic to inhalants, the vast majority were also allergic to egg. So egg allergy is associated with a higher risk of being inhalant-allergic, which in turn is associated with a high risk of subsequently being asthmatic (13).

1.9   Conclusions

The factors affecting the early life origins of food allergy are beginning to be clarified. It is important to understand that *in utero* sensitisation is but the first step in a whole atopic march. The hygiene hypothesis is, at the moment, the most credible underlying explanation for a lot of what is being observed in affluent western communities. There are several intervention studies planned over the next few years that are going to address that hypothesis to see whether it is possible to switch off some diseases, so that in the future it might be possible to prevent, even if not cure, allergy in an appreciable percentage of cases.

1.10   References

1.   Jones C., Kilburn S., Warner J.A., Warner J.O. Intrauterine environment and fetal allergic sensitisation. *Clinical and Experimental Allergy*, 1998, 28, 655-9.

2.   Holloway J.A., Warner J.O., Vance G.H.S., *et al.* Detection of house dust mite allergen in amniotic fluid and umbilical cord blood. *Lancet*, 2000, 356, 1900-2.

3.   Jones C.A., Vance G.H.S., Power L.L., Pender L.F., MacDonald T.T., Warner J.O. Costimulatory molecules in the developing human gastrointestinal tract; a pathway for foetal allergen priming. *Journal of Allergy and Clinical Immunology*, 2001, 108, 235-41.

4.   Casas R., Björksten B. Detection of Fel d 1-immunoglobulin G immune complexes in cord blood and sera from allergic and non-allergic mothers. *Pediatr. Allergy Immunol.*, 2001, 12, 59-64.

5.   Jenmalm M.C., Björksten B. Cord blood levels if Immunoglobulin G subclass antibodies to food and inhalant allergens in relation to maternal atopy and the development of atopic disease during the first 8 years of life. *Clinical and Experimental Allergy*, 2000, 30, 1, 34-40.

6.   Ruiz R.G.G., Kemeney D.M., Price J.F. Higher risk of infantile atopic dermatitis from maternal than from paternal atopy. *Clin. Exp. Allergy*, 1992, 22, 762-6.

7.   Jones C.A., Warner J.A., Warner J.O. Fetal swallowing of IgE. *Lancet*, 1998, 351, 1859.

8.   Jones C.A., Holloway J.A., Popplewell E.J., Diaper N.D., Legg J.P., Holloway J.W., Vance G.H.S., Grimshaw K.E.C., Warner J.A., Warner J.O. Reduced soluble CD14 levels in amniotic fluid and breast milk are associated with the subsequent development of eczema. *Journal of Allergy and Clinical Immunology*, In press.

9.   Warner J.A., Miles E.A., Jones A.C., *et al*. Is deficiency of interferon gamma production by allergen triggered cord blood cells a predictor of atopic eczema? *Clin. Exp. Allergy*, 1994, 24, 423-30.

10.   Warner J.O. Worldwide variations in the prevalence of atopic symptoms: what does it all mean? *Thorax*, 1999, 54 (suppl. 2), 46-51.

11.   Aaby P., Shaheen C.B., Heyes C.B., et al. Early BCG vaccination and reduction in atopy in Guinea-Bissau. *Clin. Exp. Allergy*, 2000, 30, 644-50.

12.   Kalliomaki M., Salminen S., Arvilommi H. Probiotics in primary prevention of atopic disease: a randomised placebo controlled trial. *Lancet*, 2001, 357, 1076-9.

13.   Warner J.O. for the ETAC™ Study Group. A double blind randomised placebo controlled trial of cetirizine in preventing the onset of asthma in children with atopic dermatitis: 18 months' treatment and 18 months' post-treatment follow-up. *Journal of Allergy and Clinical Immunology*, 2001, 108, 929-37.

2. PROTEINS AS ALLERGENS: A TOXICOLOGICAL PERSPECTIVE

Dr Rebecca J. Dearman and Professor Ian Kimber

2.1 Introduction

This paper focuses on proteins as allergens, particularly from a toxicological perspective.

From a toxicological or safety assessment perspective, one of the most important potential adverse health effects that may arise from the introduction of a novel food, or a novel protein engineered into a food, such as by genetic modification, is that a new allergen may be introduced into the food chain. In order to prevent this, there need to be strategies in place that are able to identify the potential of proteins to cause allergenicity.

In addition, if the elements of certain proteins, their structure, or biochemistry, which causes them to be potent allergens, can be identified, it might be possible in the future to target and remove those features and, therefore, allow the development of hypo-allergenic foods.

2.2 Proteins as Allergens

Every day we are exposed, through our gastrointestinal tracts, and also through our respiratory tracts, to a vast array of foreign proteins. Many of these hundreds of proteins are recognised as foreign by the immune system. They are immunogenic, but they are not allergenic. They fail to cause the quality of immune response, the IgE, Th-2-type inducing response, that will give rise to allergy.

At the other end of the spectrum of allergenicity there are proteins from foodstuffs such as a peanut that are highly allergenic. Clearly, there is a wide spectrum of allergenicity and what is required is to be able to identify the proteins at the top end of this spectrum - those proteins that are capable of causing allergic responses.

While it is recognised that a selected range of foodstuffs and selected proteins within those foodstuffs tend to be the main causes of the IgE-mediated allergic responses, there is also a large element of individual susceptibility; not all individuals go on to develop IgE allergic responses to those particular foodstuffs.

2.3   Susceptibility to Allergy

There is a whole variety of factors that may impact on individual susceptibility to the development of allergy, including dietary habits, age of first exposure, pattern and duration of exposure, and environmental factors, a genetic predisposition to atopy, maternal atopic status, health status and immune status. The health of the gastrointestinal tract is particularly important. If the gastrointestinal tract is damaged, and there is an on-going inflammatory response, that will impact on allergic sensitisation. In a situation where a material that is capable of causing allergy comes into contact with a susceptible individual, it is possible that allergic sensitisation may occur, and therefore the development of allergy.

2.4   Safety Assessment of Proteins

The first stage of the toxicological process, the safety assessment process, must be to define the intrinsic hazard of protein allergens, i.e. which proteins are capable of causing IgE responses and then overlaying that intrinsic hazard is the range of individual susceptibilities to allergy. The full risk assessment process will take into account intrinsic hazard and all of these possible factors that may impact on an individual's potential to develop allergy.

It is IgE that is the principal biological effector of most allergic reactions, particularly food allergic reactions, and also reactions associated with respiratory allergy, asthma, and rhinitis. In susceptible individuals, exposure to an allergenic protein will cause the induction of IgE antibody. That IgE antibody will distribute systemically throughout the body, so that individual is now sensitised. The IgE antibody binds to the surface of mast cells, which are found in all vascularised tissue, including the gastrointestinal and respiratory tracts. Once mast cells have IgE bound to them, the individual is sensitised and is susceptible to a second challenge with that same allergen. The second challenge with the same allergen causes cross-linking of the IgE molecules on the surface of the mast cell, and that triggers the mast cell to degranulate, releasing various vasoactive mediators, which cause the inflammatory response.

2.5   Mechanism of Immune Response

The balance between the negative signals from T helper-1 cells (Th-1) and the positive signals from T helper-2 cells (Th-2) determines whether an IgE antibody response is produced. B cells are the cells of the immune system that are responsible for dividing and differentiating into plasma cells which produce antibody, including IgE antibody. B cells produce IgE antibody, but require input from T helper cells. The balance between the inhibitory signals from Th-1, particularly interferon -gamma, and the promoting signals from Th-2 cells, particularly interleukin 4 (IL-4) and interleukin 13 (IL-13), will determine whether an IgE-mediated allergic response is produced.

2.6   Features of Proteins that are Important in their Allergenicity

There are a number of characteristics of proteins that seem to be associated with the ability of some proteins to be very potent allergens.

The following factors are all important:

● Size. This does appear to be of some importance in the ability of proteins to stimulate allergic IgE responses; this is because the mast cell has a minimum size requirement for a protein to cross-link those IgE molecules on the surface of the cell. That minimum size is about 30 amino acids or a molecular weight of 3 kD. If a protein is less than 3 kD, it is unable to cross-link IgE molecules on the surface of the mast cell, and will be unable to cause the clinical manifestations of the allergic response. Indeed, many of the very potent allergens that we know (the peanut allergens ara h 1 and ara h 2 or the soya bean allergen Gly m Bd) have very many allergenic epitopes, and very many IgE binding sites, allowing them to cross-link IgE on the surface of mast cells very efficiently. This probably contributes to their ability to be potent allergens.

- Abundance. Protein allergens are present in relatively large amounts in their originating foodstuff. Commonly, the major allergens of allergenic foods appear as a very large proportion of the total protein in that foodstuff. For example, ovalbumin and ovomucoid, which are the two major allergens in chicken egg, are present at 54% and 11% of the total protein, respectively. Clearly, the immune system is being exposed to very large amounts of this protein in each case.

- Stability. A very common feature of protein allergens is their stability to heat, and to the processing of the food itself, and this results in consumers being exposed to these materials in their native state in higher concentrations than if these materials were very labile. They are also resistant to proteolytic digestion - i.e., resistant to those enzymes that are present in the gastrointestinal tract that normally act to reduce large proteins to smaller proteins that our digestive system can handle.

In a recently published paper (1) pepsin was used to test the resistance to proteolytic digestion of allergenic and other common plant-derived proteins. Pepsin is the main proteolytic enzyme in gastric fluid. The common plant proteins were very labile, and disappeared very quickly; within 15 seconds, most of these proteins had been completely digested into polypeptides. The major protein allergens tested included ovalbumin, ovomucoid, β-lactoglobulin, ara h 1, peanut lectin, and Sin a 1. These proteins were very stable to proteolytic digestion. (See Table 2.1)

TABLE 2.I
Stability of major protein allergens to proteolytic digestion (1)

Protein allergen	Time taken for digestion by pepsin
Ovalbumin	60 min
Ovomucoid	8 min
β-Lactoglobulin	60 min
ara h 1	60 min
Peanut lectin	8 min
Sin a 1	60 min

Some of the protein allergens, such as ovalbumin, were still more or less intact after exposure to proteolytic enzymes for up to an hour, demonstrating that the major allergens appear to be very stable and very resistant to digestion by proteolytic enzymes. This is an important factor when considering exposure of consumers to these materials. These allergenic proteins survive in the gastrointestinal tract and there is therefore going to be more opportunity for them to interact with the immune system, and more opportunity for them to cause an immune response.

Exposure to allergenic proteins should not be considered just in terms of exposure to the total amount of protein; it is believed that these proteins are resistant not only to extracellular digestion, such as occurs in the gastrointestinal tract by pepsin, but also to the intracellular processing that is performed by the cells of the immune system - the antigen processing cells.

2.7   *In vivo* Allergen Recognition

The process of recognition of protein allergen begins with antigen processing cells internalising protein. There are pockets of enzymes within antigen presenting cells that focus the allergens, process them and then subsequently present them on the surface of the antigen processing cell so that the other cells of the immune system can interact with them and initiate the immune response. The antigen processing cells are important cells in directing whether T cells are going to develop down the Th-1 or a Th-2 pathway. B cells and T cells do not recognise the whole protein, so the function of the antigen processing cells to present only a portion of the protein on the surface of the cell is key in initiating an immune response.

It is possible that the stability of protein allergens to digestion by pepsin is actually an indication of their stability to digestion internally by antigen processing cells, and this stability will impact on the nature, and the quality of immune response that is then generated. The more stable allergens will induce Th-2 type responses, rather than Th-1 type responses.

2.8   Biological Activity of Allergens

Another common feature of protein allergens is their biological activity: common inhalation allergens include the major allergen of house dust mite, Der p 1, which is a cysteine protease; the major allergen of bee venom is phopholipase; many of the enzymes used in detergents are utilized because of their proteolytic activity, and are also very potent respiratory allergens - there was a large increase in asthma in the detergent industry as a result of the initial introduction of proteolytic enzymes; the major allergens in soft fruits, such as Mal d 1 and Pru p1, are lipid transfer proteins; and the allergen hevamine in latex is a chitinase.

There are many potential modes of action to explain why proteins that have enzymic activity might induce the quality of IgE and Th-2 type immune responses that then lead to the development of allergy. It has been demonstrated that extracellular enzymic activity stimulates Th-2, rather than Th-1 cells. In addition, the proteolytic activity of the protein allergen will augment cell permeability.

Extracellular enzymic activity will induce production of inflammatory cytokines, and it has long been known that inflammation will facilitate all aspects of the immune response, including the development of allergies. The enzymic activity of certain allergens, particularly phospholipase A2, causes direct stimulation of those key mediators of the allergic response, the mast cells. Also, the design of enzymes, and their function, cause them to be relatively stable in hostile environments. So this all contributes to the general stability to proteolytic digestion.

2.9   Glycosylation

Another common feature of protein allergens is that they tend to be glycosylated; they have sugar residues. For some time now it has been known that many of the IgE epitopes on protein allergens are glycosylated. In addition, the fact that these molecules have sugar residues will impact on their processing. This is because the allergen processing cells have receptors for sugars on their cell surface. When they bind to proteins that contain sugar residues through these sugar receptors, these proteins are internalised 100- to 1,000-fold more quickly than proteins that do not contain such sugar residues. Therefore, glycosylation augments the ability of these proteins to interact with the immune system.

2.9.1   Effect of glycosylation on allergenicity of protein

An important question is whether glycosylation impacts on the ability of the protein to induce IgE responses? This has been investigated using lactoferrin, which is a member of the iron binding transferrin family of proteins found in exocrine secretions. The lactoferrin was used in two forms: one, a recombinant form produced by a fungus, *Aspergillus niger var awamori*, and the other the native form of the protein. The two proteins are almost identical in terms of 3D structure and amino acid sequence. The only small difference is at the n-terminal end of the proteins.

To all intents and purposes, these materials are identical in terms of amino acid sequence and 3D structure. There are two iron binding sites, and, as its name suggests, lactoferrin is an iron binding protein. The recombinant and the native forms of lactoferrin differ with respect to their sugar residues, enabling investigation into how these two forms of material differ in their ability to induce IgE responses in an experimental model (2).

BALB/c strain mice of a high IgE-responder atopic-type strain were exposed by intraperitoneal injection on day 0 and 7 to either the native or the recombinant form of lactoferrin. On day 14, serum was taken from the mice, and analysed for IgG antibody production, using enzyme-linked immunobsorbent assay (ELISA). The IgE response was analysed by passive cutaneous anaphylaxis (PCA) assay.

2.9.1.1 Results

The native lactoferrin was shown to be more immunogenic, and able to induce much higher IgG responses than the recombinant form. In addition, only the native form of lactoferrin was able to induce IgE. Therefore, differences in glycosylation were able to induce different IgE antibody patterns.

The immunological identity of the native lactoferrin and the recombinant lactoferrin was investigated with respect to IgG and IgE binding activity, and they are completely identical; there are identical IgG and IgE-binding epitopes on both molecules. So the ability of IgE to bind to both forms of lactoferrin is not in question; it is the influence of sugar residues on the ability of these materials actually to induce IgE responses that is of central importance. It is believed that it is at the stage of antigen processing and handling that glycosylation plays such an important role.

2.10 Immunogenicity

The final parameter that is important is immunogenicity. In some respects, this is obvious; in order to have an allergic response there must be recognition by the immune system, and the immune system must recognise that material as foreign. However, many foreign materials are recognised as immunogenic and are capable of inducing an IgG antibody response. In order to be allergenic, however, protein allergens must be able to induce the type of immune response necessary to cause the induction of IgE, i.e. be capable of stimulating Th-2 cells, causing the balance between interferon - gamma and IL-4 to shift towards IL-4, resulting in the induction of IgE. To date, there is very little known about why certain protein allergens are able to stimulate Th-2 cells preferentially.

None of the features of potent allergens that are related to their ability to be allergenic is necessarily exclusive to allergens. For example, whereas most allergens are indeed stable, and stability does appear to be a property of most allergens, not all stable proteins are allergens; there are some stable proteins that are incapable of stimulating the quality of immune response necessary to cause the IgE-mediated response. It is a combination of these features that confers on a certain protein the ability to cause allergy and any one feature in isolation is not enough to cause that protein to be an allergen.

2.11 Testing Allergenicity of Proteins in an Animal Model

Proteins vary in their ability to induce IgE responses in an animal model of allergenicity. The intrinsic capability of certain protein allergens to cause IgE responses can be tested by exposing animals to the native protein, without using adjuvant, or manipulating the particular quality of immune response. Thus, it is the inherent capacity of the allergen to induce IgE responses that is under investigation. It is possible to investigate the immunogenicity of proteins with respect to their IgG antibody production and their allergenicity with respect to their ability to induce IgE.

2.11.1 Testing IgE response

Ovalbumin - a major egg protein allergen, peanut agglutinin (a minor peanut allergen), and purified protein extract (PPE) from the potato - a foodstuff that is rarely associated with allergenicity were tested for their ability to induce an IgE response in an animal model as a function of time following exposure to allergen: 14 days, 28 days, and 42 days (3). BALB/c strain mice received a systemic (intra peritoneal) injection of protein on days 0 and 7.

2.11.1.1 Results

There was a marked IgE antibody response to ovalbumin and peanut agglutinin in the majority of animals throughout the time course examined, whereas the potato protein, which is not particularly associated with allergy, had a very reduced potential to induce IgE antibody responses.

The different ability of these proteins to induce IgE antibody responses in BALB/c strain mice appears to be fairly reproducible. In three different experiments where animals were exposed to the potato protein, ovalbumin, or peanut allergen, peanut and ovalbumin induced very high IgE antibody responses in the three experiments, whereas, again, the potato extract failed to induce a high IgE response.

In these experiments, the intrinsic capacity of a protein to cause IgE induction was being investigated. The protein was delivered by a very immunogenic route in order to give it the best chance possible of inducing IgE in order to differentiate between those proteins that have a high potential to induce IgE and those proteins that have little or no potential to cause IgE. Clearly, the route of exposure will impact on the ability of proteins to induce IgE antibody responses.

2.11.2 Testing route of exposure to protein allergen on IgG and IgE production

An important question is: is protein allergenicity dependent on the route of exposure? The following study compared the systemic route of exposure with intraperitoneal (IP) exposure and the oral route (4).

BALB/c strain mice were exposed to ovalbumin, peanut allergen and potato protein by daily oral gavage, and, at 28 or 42 days after exposure, the serum was analysed for IgE and IgG antibody production.

2.11.2.1 Results

Results showed that, 28 or 42 days after exposure, the peanut allergen elicited a very strong IgG antibody response in all the animals. Ovalbumin was less immunogenic, with some animals failing to respond. Twenty-eight days after exposure to the potato protein there was very little IgG response; however, after 42 days, there was the beginning of an induction of an IgG response.

In summary following oral exposure, peanut allergen was very immunogenic - ovalbumin somewhat less so. The potato protein was also immunogenic 42 days after exposure.

The peanut allergen was able to induce a very strong IgE antibody response, whereas exposure to the less potent allergen ovalbumin by the oral route did not induce a strong IgE antibody response. Low titre IgE antibody responses were also observed following treatment with the potato protein.

If these results are compared with those achieved by delivering allergens by the intraperitoneal route, the response in terms of IgG generation is similar. Both the common allergens, peanut and ovalbumin, induced strong IgG antibody responses, and the potato protein was also clearly very immunogenic - recognised as foreign by the immune system.

Following systemic (intraperitoneal) exposure, however, it is possible to observe a very good distinction between the ability of these proteins to induce IgE; only the peanut and ovalbumin allergens induced high IgE levels, and the potato protein failed to induce detectable IgE antibody under most dosing regimes.

Therefore, by exposing animals through a route that is highly immunogenic, equivalent doses in terms of immunogenicity can be administered, and a very clear distinction in terms of the ability of these proteins to induce IgE (to cause allergenicity) has been demonstrated.

In terms of our ability to characterise the intrinsic hazard of a material's ability to induce IgE and, therefore, allergenicity, the toxicological problems associated with the introduction of novel proteins are beginning to be unravelled. The challenge for the future is being able to apply that information about the intrinsic hazard of a material to the risk assessment process and also understanding the particular properties of protein allergens that cause them to induce IgE and allergic reactions.

2.12   References

1.   Astwood J.D., Leach J.N., Fuchs R.L. Stability of food allergens to digestion *in vitro. Nature Biotech.*, 1996, 14, 1269-73.

2.   Dearman R.J., Headon D.R., Kimber I. Lack of allergenicity of recombinant lactoferrin: role of glycosylation. *Tox. Sci.*, 2000, 54, 248.

3.   Dearman R.J., Kimber I. Determination of protein allergenicity: studies in mice. *Tox. Letters*, 2001, 120, 181-6.

4. Dearman R.J., Caddicle H., Stone S., Basketter D.A., Kimber I. Characterisation of antibody responses induced in rodents by exposure to food proteins: influence of route of exposure. *Toxicology*, 2001, 167, 217-31.

3. MILK ALLERGY

Jean-Michel Wal

3.1 Introduction

This paper provides an overview on milk allergy, and the structure and function of milk allergens in particular.

Milk allergy is one of the major allergies in infants (1). However, there are no definite epidemiological data on the prevalence of milk allergy in Western countries, which is believed to be 2-3% of the general population in children under 2 years of age. These figures concern true milk allergy, that is to say, the IgE-mediated form.

Children are generally considered to outgrow milk allergy spontaneously after 3 years. However, milk allergy may persist and occur in adults. According to a recent study by Wüthrich in Switzerland, females appear to be more affected by milk allergy than men (2).

3.2 Milk Proteins

As in any food, the components that are responsible for allergy in milk are the proteins. Indeed, most milk proteins are potential allergens whatever their structure, chemical properties, i.e. stability, function or biological activity may be.

Milk protein is qualitatively similar in different ruminant species (see Table 3.I) but concentrations of the different proteins may vary. They may also vary depending on the stage of lactation. It is worthy of note that human milk does not contain β-lactoglobulin.

Milk contains between 30 and 35 g protein per litre (See Table 3.II), and the action of the proteolytic enzyme chymosin produces two protein fractions – whey, which contains 20% of the protein, and the coagulum, which contains a casein fraction accounting for about 80% of the milk protein.

TABLE 3.I
Protein composition of milk of different mammalian species
(g/l)

	Cow	Goat	Ewe	Mare
Whole casein	28-30	25-30	50-60	13
αS casein	14	2-6	25	
β casein	11	18	25	
κ casein	4	4	10	
Whey proteins	6	4	9	13

TABLE 3.II
Characteristics of the major milk proteins (adapted from (4))

	Whey proteins	Concentration (g/l)	Mol. weight (kDa)	No. A.A.	pI
	β-lactoglobulin	3-4	18.3	162	5.3
20%	α-lactalbumin	1-1.5	14.2	123	4.8
~5 g/l	Lactoferrin	0.09	80	703	8.7
	Proteose-peptones	0.5-1.5			
	Albumin	0.1-0.4	67	582	4.9-5.1

	Caseins	Concentration (g/l)	Mol. weight (kDa)	No. A.A.	pI
	αS1-casein	12-15	23.6	199	4.9-5
80%	αS2-casein	3-4	25.2	207	5.2-5.4
~30 g/l	β-casein	9-11	24	209	5.1-5.4
	κ-casein	3-4	19	169	5.4-5.6
	γ-casein	1.5-2	11.5-20.5	102-181	5.5-6.7

Whey essentially contains globular proteins. The main ones, i.e. β-lactoglobulin and α-lactalbumin, are synthesized in the mammary gland while others come from blood.

The casein fraction comprises four main different proteins coded for by different genes, namely, α-S1, α-S2, β- and κ-casein. Each of these caseins represents a well-defined chemical compound. The different caseins associate with each other in solution, and cross-link to form complexes, then ordered aggregates called micelles that are in suspension in the aqueous phase of the lactoserum. The proportion of the different caseins in the micelles is quite stable ca 34, 36, 10 and 13% of the whole casein for β-, α-S1, α-S2 and κ-casein, respectively. Moreover, the distribution of proteins is not uniform within the micelles, which are formed of a central hydrophobic part and a peripheral hydrophilic layer where the calcium binding sites are exposed, in accordance with the calcium-binding and transfer properties of caseins (3).

3.3 Allergenicity of Milk Proteins

The milk proteins are all potential allergens. A global approach to defining the antigen and allergen repertoire of milk can be realised by two-dimensional electrophoresis, which separates the proteins, which are then characterised by their molecular mass and isoelectric point. An immunoblot is then performed on nitrocellulose membranes using sera of either control or allergic patients in order to reveal which proteins are antigenic, or allergenic.

Allergenic proteins are recognised by IgE, and antigenic proteins are recognised by IgG. Some proteins are recognised by both IgE and IgG, sometimes with differing intensity. Some other proteins are recognised only by IgE and either not recognised by IgG at all, or to a much lesser extent (5).

More specific approaches using isolated and purified protein for direct and competitive inhibition ELISA test with human allergic patients' serum have also been performed (6). Figure 3.1 shows the individual patterns of IgE specificity to five milk proteins in 20 patients. The striking point is the great variability of IgE response, in terms of both specificity, i.e. the number of allergens that were recognised, and intensities, i.e., the height of the different peak of response. Particular attention should be paid to patient number 7, who had clinical symptoms of milk allergy but negative radioallergosorbent

(RAST) test and was sensitised only to lactoferrin, which is present in milk at very low concentration; lactoferrin is a minor protein in terms of its concentration in food, it is nevertheless an important allergen, and sometimes it is the only one to be identified as provoking an allergic reaction (7).

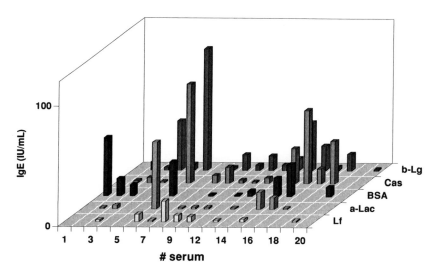

Legend: b-Lg: β-lactoglobulin
Cas: whole casein fraction
BSA: bovine serum albumin
a-Lac: α-lactalbumin
Lf: lactoferrin

Fig. 3.1. EIA of specific IgE to five milk proteins (20 patients)

3.4   Trends in Sensitisation to Milk Proteins

In terms of prevalence of sensitivity to milk in a large population of allergic patients, what is interesting is that the pattern of sensitisation is not the same now as it was in 1990. Figure 3.2 demonstrates the results of a study of a population of about 80 patients with clinical symptoms of milk allergy in the past 2 years as compared with results from 10 years ago in a similar population. There is still a high degree of polysensitisation, but the relative proportion of allergy is less homogeneous nowadays, with the majority of patients demonstrating sensitisation to two allergens.

It has also been observed that the prevalence of sensitisation to casein has much increased in comparison with what was observed in 1990. At this time, prevalence of β-lactoglobulin and casein allergy was at the same level; indeed, a few years before this study, β-lactoglobulin was considered as the only milk allergen. Its importance as an allergen has now decreased, as well as that of α-lactalbumin and whey protein, generally speaking, with the exception of bovine serum albumin (BSA), but BSA does not appear to be a relevant marker for cows' milk allergy (CMA).

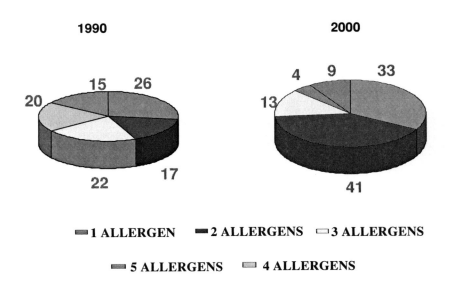

Fig. 3.2. Mono- or multi-sensitivities to milk allergens (ca. 80 patients)

3.5 Structural and Functional Properties of Milk Allergens

3.5.1 Caseins

The main allergens (α-S1, α-S2, β- and kappa-casein), which form the whole casein fraction, are coded by four different genes and have very little amino acid sequence homology; however, they display common features. Caseins are often considered to be poorly immunogenic because of a flexible, non-compact structure. Moreover, caseins are not significantly affected by severe heat treatments but are very susceptible to all proteases and exopeptidases, and are rapidly and extensively hydrolysed by digestive enzymes. They share a dipolar type structure with amphipathic properties. Both αS- and β-caseins are calcium-sensitive. They are phosphorylated proteins with a loose tertiary highly hydrated structure. The molecules have a globular hydrophobic domain and a highly solvated and charged domain. The striking common characteristic is the homology in the acidic peptide sequence, containing a cluster of phosphoseryl residues. There are two major sites of phosphorylation in αS-2-caseins, and one in αS1- and β-casein as shown in Figs 3.3, 3.4 and 3.5 (8-10). Several isoforms have been described for each casein. These variants are characterised by point substitution of amino acids or by deletion of peptidic fragments of varying size or by post-translational modifications. All these modifications may affect the allergenicity.

Table 3.III illustrates the high sequence homology of caseins in cows' milk, and in the milk of other mammalian species. This high sequence homology varies between 80 and more than 90% homology. As a consequence, IgE cross-reactivity between ewes', goats' and cows' milk casein occurs in most patients with cows' milk allergy, and the IgE response to goats' and ewes' milk is similar to or even sometimes higher than that to cows' milk. Between 15 and 20% of patients allergic to cows' milk casein have a similar or even higher response to caseins from ewes' or goats' milk. This casts much doubt on the suggestion that allergic reaction can be avoided by using milk from species other than cow, and on the continuing use of goats' milk as a substitute for cows' milk (11).

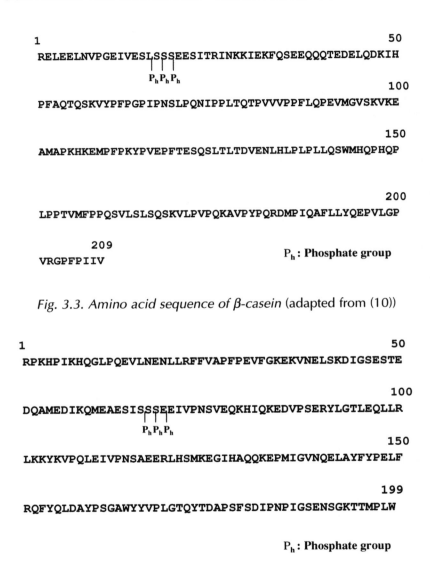

1   50
RELEELNVPGEIVESLSSSEESITRINKKIEKFQSEEQQQTEDELQDKIH
            P_h P_h P_h
                                                    100
PFAQTQSKVYPFPGPIPNSLPQNIPPLTQTPVVVPPFLQPEVMGVSKVKE

                                                    150
AMAPKHKEMPFPKYPVEPFTESQSLTLTDVENLHLPLPLLQSWMHQPHQP

                                                    200
LPPTVMFPPQSVLSLSQSKVLPVPQKAVPYPQRDMPIQAFLLYQEPVLGP

        209
VRGPFPIIV   P_h : Phosphate group

Fig. 3.3. Amino acid sequence of β-casein (adapted from (10))

1   50
RPKHPIKHQGLPQEVLNENLLRFFVAPFPEVFGKEKVNELSKDIGSESTE

                                                    100
DQAMEDIKQMEAESISSSEEIVPNSVEQKHIQKEDVPSERYLGTLEQLLR
            P_h P_h P_h
                                                    150
LKKYKVPQLEIVPNSAEERLHSMKEGIHAQQKEPMIGVNQELAYFYPELF

                                                    199
RQFYQLDAYPSGAWYYVPLGTQYTDAPSFSDIPNPIGSENSGKTTMPLW

                              P_h : Phosphate group

Fig. 3.4. Amino acid sequence of αS1-casein (adapted from (8))

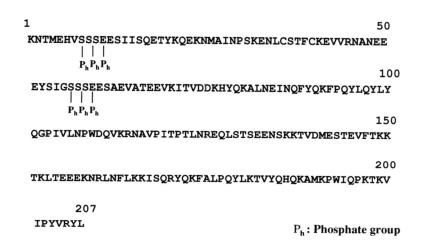

Fig. 3.5. Amino acid sequence of αS2-casein (adapted from (9))

3.5.2 Whey proteins

Whey proteins have a compact and rigid structure that is stabilised by disulfide bonds. Lactoferrin is a large protein corresponding to serum transferrin and is a carrier for iron. It is present in only trace amounts in whey.

Interestingly, proteose peptones that are present in whey are short peptidic fragments derived from splitting of the N-terminal moiety of gamma-casein, which is naturally formed after hydrolysis of β-casein by endogenous enzymes, such as plasmin, during milk conservation. These proteose-peptones are soluble in whey; they are casein fragments but they are present in whey, which means that allergic patients sensitised to caseins but with a negative RAST to whey protein may suffer an allergic reaction after ingestion of whey protein hydrolysate formula because it may contain those soluble fragments of casein.

TABLE 3.III
Sequence homologies between milk proteins in different species (%)

β Lactoglobulin

	Cow	Goat	Ewe	Mare
cow	100	96	96	45-60
goat		100	99	45-60
ewe			100	45-60
mare				100

α Lactalbumin

	Cow	Goat	Ewe	Mare
cow	100	95	94	75
goat		100	99	73
ewe			100	73
mare				100

Casein	Cow/Ewe	Cow/Goat	Ewe/Goat
αS1	89	87	97
αS2	89	88	98
β	90	90	99
κ	84	85	95

3.5.3 β-Lactoglobulin

Beta-lactoglobulin occurs naturally in the form of 36 kD dimer. Each subunit corresponds to a 162-residue polypeptide. The molecules possess two disulfide bridges and one free cysteine. The structure is responsible for the resistance of β-lactoglobulin to acid hydrolysis and to digestion by protease, as well as for interaction with casein during heat treatments.

There are two main isoforms of β-lactoglobulin, i.e. the genetic variants A and B, which differ only in two point mutations on residues 64 and 118, with aspartic and valine residues on variant A, and glycine and analine residues in variant B.

The fact that β-lactoglobulin is very stable and resistant to hydrolysis by digestive enzyme is not the unique reason for its allergenicity. Figure 3.6 represents the results of competitive inhibition RAST test or EAST test using native β-lactoglobulin, or denatured β-lactoglobulin, or hydrolysed β-lactoglobulin (12). All the inhibition curves obtained with the sera of allergic patients are identical, which means that, even when the secondary structure is lost, or even when the β-lactoglobulin is split using cyanogen bromide, the allergenicity of the molecule remains intact, so there is possibly a correlation between stability of a protein and allergenicity, whereas there is no strict mechanistic relationship.

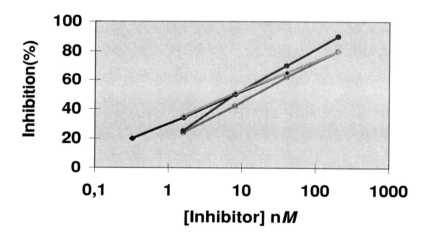

Fig. 3.6. Immunoreactivity (i.e. IgE binding capacity) of native, denatured and hydrolysed β-lactoglobulin

As is the case for casein, there is a great homology with β-lactoglobulin of cows' milk and that in other ruminants. Table 3.III shows that, with mares' milk, the homology is higher than 50%.

The tertiary structure of β-lactoglobulin is known. It belongs to the lipocalin family, which comprises many allergens and has a very conserved tertiary structure called β-barrel structures formed by arrangements of eight to ten antiparallel β-sheets. These proteins are carriers for hydrophobic ligands, and β-lactoglobulin is thought to be a retinol binding protein.

3.5.4   α-Lactalbumin

Another major whey protein is α-lactalbumin, which is a monomeric globular protein of 123 amino acid residues and four disulfide bridges. It possesses a high affinity binding site for calcium and this bond stabilises the structure. It is a regulatory component of the enzymatic system of galactosyl transferase that is responsible for the synthesis of lactose. Its structure is very close to that of egg white lysozyme. The amino acid sequence of α-lactalbumin from cows' milk also shows a great homology with α-lactalbumin from other mammalians' milk (see Table 3.III). This homology extends to human milk α-lactalbumin, since 74% of the residues are identical, and another 6% are chemically similar. In some domains, including the C-terminal fragment, the homology between bovine and human α-lactalbumin is higher than 80% to 85%.

3.6   Conclusion

The main characteristics of milk allergens are summarised below:

- Multiplicity and diversity of allergens in milk and variability and heterogeneity of human-specific IgE response.
- No protein or structure can be described as being intrinsically allergenic.
- No single epitope is responsible for a whole protein's allergenicity; most protein molecules contain several fragments with IgE-binding capacity.
- Allergenic epitopes may be different from antigenic epitopes as predicted by structure analysis/animal models.

– Existence of both conformational and linear epitopes short fragments (12-14 amino acid residues), located in hydrophobic/denatured parts of the protein molecule and unmasked after digestion.

All these factors lead back to the conclusion that great care should be taken when using peptidic fragments in hydrolysed aids for allergic infants or for desensitisation.

3.7   References

1.   Host A. Cow's milk protein allergy and intolerance in infancy. *Pediatr. Allergy Immunol.*, 1994, 5 (suppl 5), 1-36.

2.   Stöger P., Wüthrich B. Type I allergy to cow milk proteins in adults. A retrospective study of 34 adult milk- and cheese-allergic patients. *Int. Arch. Allergy Immunol.*, 1993, 102, 399-407.

3.   Swaisgood H.E. Chemistry of milk protein, in *Developments in dairy chemistry-1 Proteins*. Ed. Fox P.F. London and New York, Applied Science Publishers. 1982, 1-60.

4.   Jost R. Physicochemical treatment of food allergens: application to cow's milk proteins. *Nestlé Nutrition Workshop Series*, 1988, 17, 187-97.

5.   Brodard V., Bernard H., Wal J.M., David B., Peltre G. Two-dimensional analysis of cow's milk allergens with the IPG-DALT technique, in *Recent Advances in Mucosal Immunology*. Eds McGhee J., Mestecky J., Tlaskalova H., Sterzl J. Series Advances in Experimental Medicine and Biology. New York, Plenum Press. 1995, 875-78.

6.   Wal J.M., Bernard H., Yvon M., Peltre G., David B., Creminon C., Frobert Y., Grassi J. Enzyme immunoassay of specific human IgE to purified cows' milk allergens. *Food Agricultural Immunol.*, 1995, 7, 175-87.

7.   Wal J.M., Bernard H., Creminon C., Hamberger C., David B., Peltre G. Cow's milk allergy: the humoral immune response to eight purified allergens, in *Recent Advances in Mucosal Immunology*. Eds McGhee J., Mestecky J., Tlaskalova H., Sterzl J. Series Advances in Experimental Medicine and Biology. New York, Plenum Press. 1995, 879-81.

8. Mercier J.C., Grosclaude F., Ribadeau-Dumas B. Structure primaire de la caseine-αS_1 bovine: séquence complète. *Eur. J. Biochem.*, 1971, 23, 41-51.

9. Brignon G., Ribadeau-Dumas B., Mercier J.C., Pelissier J.P., Das B.C. Complete amino acid sequence of bovine αS_2-casein. FEBS Lett, 1977, 76, 274-79.

10. Ribadeau-Dumas B., Brignon G., Grosclaude F., Mercier J.C. Structure primaire de la caséine-β bovine. Séquence complète. *Eur. J. Biochem.*, 1972, 25, 505-14.

11. Bernard H.B., Creminon C., Negroni L., Peltre G., Wal J.M. IgE cross reactivity of caseins from different species in allergic humans to cow's milk proteins. *Food Agric. Immunol.*, 1999, 11, 101-11.

12. Selo I., Negroni L., Yvon M., Peltre G., Wal J.M. Allergy to bovine β-lactoglobulin: specificity of human IgE using CNBr derived peptides. *Int. Archs. Allergy Immunol.*, 1998, 117, 20-28.

4. PEANUT AND TREE NUT ALLERGY: WHY SO SCARY?

Dr Jonathan Hourihane

4.1   Introduction

Bonfire night is a very risky time for children with nut allergy because, from childhood memories, it is a time when monkey nuts, which are what peanuts are in their unroasted form, come out in the shops, along with the other treats of autumn harvest. Could monkey nuts be considered as a treat? Should a child with an allergy come to the door trick or treating, would the average person, if they asked the child "Do you have peanut allergy?" and they said "No" be confident enough to give them peanuts? The point is that allergy to peanut impacts on all parts of a child's life.

Data from Britain and the United States over several different areas of peanut allergy show that the prevalence may be increasing, but it was shown a few years ago (1) – and there have been further confirmatory studies (2,3) - that peanut allergy and tree nut allergy may be outgrown, although this may be less likely for tree nut allergy.

It is clear that mild cases of peanut allergy can progress and it is possible that there is no such thing as a low-risk peanut allergic child. That is quite a significant statement because many children can go for a long time between reactions, but, in a fair proportion of children, a severe reaction will eventually occur. These are some of the most important data to be published recently (3).

The fact remains that peanuts and tree nuts remain the leading cause of food-related anaphylaxis; even though only 1% or less of the population is allergic to them, compared with milk, or egg, where in some groups 3% to 4% of children are allergic. Thus they feature disproportionately in the incidence of food-induced anaphylaxis, and this situation persists despite the efforts of the medical profession.

It is very hard to avoid peanuts in the diet, and that is likely to be a major focus for people in the food industry. Unfortunately, there is still no treatment for peanut allergy.

4.2   Allergen Sensitivity

An example of how little allergen is required to cause an allergic reaction can be demonstrated by a boy who is allergic to peanuts. The boy kissed his mother after she had eaten some trout, which he had never eaten. He did not know he was allergic to trout, and, after kissing her on the lips after she had eaten some trout, he developed a massively swollen lip. Erythema developed around his chin and then his eyes became itchy. This was not due to a generalised reaction; he rubbed his eye after rubbing his mouth because it was itching. So there was a very low level of contact with the allergen as a result of kissing somebody, which caused localised reactions.

There have also been reports of significant generalised reactions after kissing - for example in a peanut-allergic person who kissed another person who had been eating peanuts (4). There are real risks related to dose and route of exposure, as has been shown in animal models (5,6). There are also risks and evaluations that need to be made with regard to people who have had mild symptoms in relation to tiny exposures. It is not difficult to deduce that, if someone who has previously had a mild reaction has a larger dose of the allergen in question orally, he/she may have a more severe reaction (3).

4.3   Prevalence of Peanut Allergy

A study conducted on the Isle of Wight in 1989 produced data to show that 0.6% of a population of 900 four-year-old children included in the study had peanut allergy and that 1.3% of them were sensitised to peanut allergy (1). Of the group of 900 children, 13 had a positive skin prick test, but only 6 were allergic.

A repeat study (7) has been done with a larger number of children, who were again followed to 4 years of age. A larger number of subjects were skin prick tested than in the previous study, and out of 1,237 who had a positive skin prick test, 41 of those tested had a positive skin prick test. The population on the Isle of Wight is relatively stable; there is not much immigration and most of the immigration involves relatively affluent people; however, 3.2% (n=41) of this group of children, on the same island as in the previous study, had a positive skin prick test to peanut, compared with the rate in the 1989 cohort reported in 1996 of 1.3%.

4.4   Resolution of Peanut Allergy

Recent research has shown that peanut allergy can be outgrown (1). This is really the only good news that peanut-allergic individuals have had in the last 5 or so years. The study conducted in Southampton (1) showed that, in children who had resolved peanut allergy (i.e. peanut allergy that appeared to have gone away), there was a clear demarcation in that nobody with resolving peanut allergy had a skin prick test of more than 6 mm. However, a couple of children in the study had skin prick tests below 6 mm and still had persistent allergy. This demonstrates that the diagnosis of peanut allergy should not be based purely on a skin prick test; there must be a challenge with the allergen in question or an unequivocal history.

In a study by Skolnick *et al.* (2), children were studied and a challenge test was performed at the start of the study. One of the criticisms of the study done by Hourihane *et al.* (1) is that challenges were not done at the beginning of the study and it was not absolutely certain that the children who had allergy that resolved had allergies in the first instance. Skolnick *et al.* (2) are certain that children that took part in their study had allergy because they had positive skin prick tests and a specific IgE level, or what would previously have been called a "RAST" level, of above 20, which is 95% predictive of disease being present, at the start of the study. Children were then retested to determine resolution of allergy with specific IgE. Ninety-seven subjects - nearly half of the original number studied - were still allergic to peanut, and 41 subjects whose IgE had fallen below 20 refused to have a challenge on follow-up. In 85 subjects, IgE levels had fallen from above 20 to below 20. Two-thirds of these children had negative challenges and one-third had a positive challenge. The resolution rate was 25% in children who were retested. If it is assumed that 56% of the children (41 in total) who refused to have peanut challenges would have had negative challenges, there could be another 23 children with resolved allergy. This might mean that peanut allergy resolves in up to 32% of cases in young children who have had challenge-proven diagnosis.

Therefore, the practice of saying that peanut allergy persists forever has got to change, particularly in regard to young children. In older children, and in children for whom tree nut allergy starts in later life, it is best to be more cautious and say that resolution is unlikely.

4.4.1 Factors important in resolution of allergy

The factors that predicted the resolution of peanut allergy in the studies by Hourihane and Skolnick (1,2) are that the children who had grown out of allergy did not have many other forms of allergy. They were not multiply atopic, with many other allergy-related manifestations. They had not had a reaction for a long time; that is unusual because, in general, a reaction is expected in a child every 2 years. Fifty per cent of the group studied by Hourihane (1) had a reaction every year; the children who demonstrated resolved food allergy had not had a reaction for up to 8 or 9 years, and this was a feature that might help suggest whose peanut allergy would resolve. The children whose peanut allergy resolved also had a low peanut-specific IgE at review (1), and this has been confirmed by Skolnick's study (2). A low skin prick test at review also confirmed it. What the study by Hourihane (1) failed to show as a predictive feature – which was a surprise, and has been confirmed – that the severity of the presenting reaction does not predict that the child will always have the allergy. The level of specific IgE at presentation is not predictive of resolving either because, in two-thirds to half of the children in Skolnick's study (2), their specific IgE had fallen from a level at which one would be very confident that a challenge would be positive to a level at which the challenge was at least worth attempting to prove resolution of allergy.

4.5   Progression of the Allergic Reaction

Mild cases of peanut allergy can progress from mild to more severe, and this is bad news for peanut-allergic individuals because many doctors, and indeed many allergists, reassure children that: "Well, you have only had a mild reaction. If you avoid peanuts you are going to be okay."

4.5.1   Factors affecting progression of the allergic reaction - asthma

One of the factors that may affect the severity of subsequent allergic reactions is the presence or absence of asthma. In a study conducted in Southampton, in a group of adults and children with mild reactions to peanut, half had asthma and half did not have asthma. In those subjects who had moderate reactions, nearly three-quarters had asthma and the remainder did not have asthma. In those children with severe reactions, nearly 75% had asthma and only 25% did not have asthma. So the presence of asthma, which is present in more than 60% of children and adults with peanut allergy, predicts severe disease. Therefore, 60% of subjects who are peanut-allergic are already at risk of having at least a moderate reaction if they have asthma.

4.5.2   Factors affecting progression of the allergic reaction – severity of first reaction

Table 4.I demonstrates how allergic reactions can change (8). The severity of the most recent reaction is given left to right, and the severity of the first reaction is given from top to bottom. In 107 patients whose first reaction was mild, only 44 had a mild reaction on their most recent reaction; two-thirds of this group had a more severe reaction subsequent to their first one. Two hundred patients had a moderate reaction on their first reaction and 85% had equivalent or more severe reactions subsequent to their first one. In the 223 subjects with a severe first reaction, three-quarters had a severe reaction again.

TABLE 4.1
Reactions can progress

		Mild	Moderate	Severe	Total
	Mild	44	29	34	107
1st reaction	Moderate	27	111	59	197
	Severe	21	29	173	223
	Total	92	169	266	527

Most recent reaction →

It can be concluded that having a severe reaction after first exposure to peanut can predict subsequent severe reactions, and there is a proportion of people with mild and moderate disease who progress to worse reactions.

The Food Allergy Network in the United States has established a nut allergy register (9). Of those registered, 25% or so had a severe first reaction to nuts. At the second reaction, the symptoms became more severe and, by the third reaction, nearly 40% of the reactions were severe. These findings may be due to increasing awareness or better appreciation of the symptoms of a reaction. This may also explain the increase in use of the epipen; better education is influencing use of treatment options, as well as a possible increase in the number of severe reactions requiring use of the epipen.

Vander Leek *et al.* (3) studied 83 children prospectively who were less than 4 years old at diagnosis of peanut allergy. This study involved a very well characterised, prospectively followed group. Of the 73% of patients who had a mild reaction, 70% of these had more reactions, 43% of which were more severe. Of the 27% of the population that had a severe first reaction, 77% had more reactions, 71% of which were severe. Therefore, severe disease predicts severe disease and mild disease can progress.

4.5.3   Factors affecting progression of the reaction – route of first exposure

Vander Leek's study (3) also showed that, if the children's first reaction was due to skin contact, 66% had more reactions and they were all more severe. So, if the first reaction to peanuts was due to skin contact, it is possible that, if the next reaction follows ingestion of peanuts, it is likely that the reaction will be more severe than the first one. However, in cases where the first reaction to peanut was following ingestion, and the reaction was only mild, two-thirds of the population had further reactions to peanut, of which 63% were more severe with further ingestion. Therefore, a history of a mild reaction in childhood is not an indication of mild reactions in the future, and it is possible that not many allergists in Britain are aware of these data.

However, one major study conducted by Ewan *et al.* (10) suggests that peanut allergy may not progress. A total of 567 subjects, adults and children, was followed prospectively. Patients were given education programmes regarding peanut allergy, and personalised care plans. Two years later, 88 patients (15%) had had a follow-up reaction of reduced severity; 62 of the 88 patients had only a mild reaction. Only 3 patients (0.5%) had a severe follow-up reaction, and 85% of patients had no further reactions. Therefore, only 15% of subjects had a repeat reaction and 85% of subjects, over variable lengths of time, had successfully avoided peanuts. Interestingly, of those who had repeat reactions, 70% were asthmatic, which is a slight over-representation. One of those patients who had a severe second reaction had had a mild first reaction and had not been prescribed any epinephrine or adrenalin rescue kits. All the repeat reactions were in adults.

The authors concluded that most children who have peanut allergy do not have severe reactions when they are re-exposed. It is hard to know whether that is due to the care plan that they had which increased their awareness, and they were taking early steps to abort the reactions with antihistamine; so whether this is real benign disease is hard to say. Because the study was not challenged-based, there was probably a proportion of people who had not had reactions over the follow-up period, who had been exposed to allergen, but their allergy had resolved. This study may have placed too much emphasis on the success of avoidance of peanut, because it is known from several studies that there is a high rate of exposure to peanuts in peanut allergy sufferers, and it is possible that in children who have not had a challenge-based diagnosis, a third of them have actually probably grown out of their peanut allergy over the follow-up period, and they have been exposed to peanut but not reacted.

The results of this study from Ewan *et al.* (10) are more reassuring for children than those from the other studies presented. There are doubtless some caveats about the design of this study, which suggest that it might be premature to stop being concerned about peanut allergy in children.

4.6   Fatal Anaphylactic Reactions to Tree Nuts and Peanuts

Peanuts and tree nuts remain the leading cause of food-related anaphylaxis. The Food Allergy Network Fatality Register showed 32 deaths from peanut anaphylaxis over 5 years (11), of which there were three deaths in children under the age of one year, including one toddler of about 12 months. The oldest patient to die from peanut anaphylaxis was 33; those whose deaths were registered were young adults and children.

The 32 deaths reported were divided into two groups: group 1 contained 21 subjects for whom there were complete data about the whole event; group 2 contained 11 subjects for whom there were incomplete data. Peanuts and tree nuts caused every single death in the group where complete data were present, and they caused 9 out of the 11 deaths in group 2. That is extraordinary, because, if one considers the number of different foods that people eat every day, or every week, and the actual number of food components that must be present, for two kinds of food to represent nearly all the cases of death from anaphylaxis over a 5-year period is an over-representation that is beyond coincidence.

In group 1, all subjects had asthma. Of the 21 subjects in group 1, 20 had had known previous reactions. So these people all knew that they had food allergy. The subject who did not have a previous reaction was the baby, who died having tasted a bit of brazil nut given to him. Ten subjects were given epinephrine, which is adrenalin. Two subjects were given epinephrine as soon as the reaction started and eight were given epinephrine later than is optimum.

In group 2, there were only four known asthmatics. There were incomplete data on the seven other subjects in this group. Nine subjects were known to have had an allergic reaction previously, and four were given epinephrine, two correctly as soon as possible, and 2 were given adrenalin late.

Therefore, of 32 subjects, 14 were given epinephrine without effect, but 10 out of the 14 were given epinephrine too late. A criticism of epinephrine is that it does not save every person suffering an anaphylactic reaction, which is true - it does not. There is no medical intervention that cures every medical problem; having your appendix out will not prevent death from complications of appendicitis, if it is left too late. If epinephrine is administered too late it is not given a chance to work. This is why allergy sufferers should be equipped with epinephrine and given the confidence and education to use it themselves, because they should be aware of the symptoms of anaphylaxis when the reaction starts, and there may not be paramedics around the corner.

4.7   Peanut Avoidance

It is very hard to avoid peanuts. In a study conducted by Niemann *et al.* (12) at the University of Lincoln in Nebraska, products that were labelled as "may contain peanuts" and were produced on shared equipment or produced in shared facilities were examined. These products are probably safe, or may be unsafe; they are labelled as possibly hazardous. Peanut was detected in all the products tested - most of the products were accurately labelled but in some the risk of reaction from peanut consumption was considerable.

A supermarket survey was undertaken in South Wales by McCabe *et al.* (13) in which 603 non-nut cereal and confectionary products were examined. Of these products, 213 products were cereals, 296 were biscuits, and 121 were confectionery. Only 15% of these products were shown to contain nuts; however, only 25% or so were guaranteed as nut-free. The "may contain nuts" label is obviously a major controversial issue and it is hard to justify the use of this form of label if it is on two-thirds of products; there must be better ways of identifying the hazard of nut contamination if at present food manufacturers can only guarantee that a quarter of products that are not meant to contain nuts actually do not contain nuts. This is a real issue that has not been adequately dealt with to completion.

4.8   Effect of Processing on Nut Allergenicity

The form in which an allergen is given to an allergic person makes a considerable difference to how the allergic reaction develops. In Westernised or Americanised societies, peanuts are commonly roasted, whereas in parts of the Far East and Africa peanuts are boiled; this may have a difference in terms of allergenicity. One study demonstrated that specific IgE binding is lower for boiled peanuts in comparison with roasted peanuts, and the same is true when comparing specific IgE binding between roasted and fried peanuts – roasted peanuts have a greater IgE response (14). Therefore, the way that peanuts are cooked may affect their relative allergicity.

Maleki *et al.* (15) studied the allergenicity of roasted and raw peanuts, and found that roasted peanuts were significantly more allergenic than raw peanuts.

4.9   Treatment for Peanut Allergy

Unfortunately, there is still no treatment for peanut allergy. There are animal models and immunisation models where reactivity can be changed, but that is only when the mice are pre-immunised before exposure. Immunisation of every child for a disease that affects only 0.5 or 1% of children cannot be justified. Measles, mumps, chicken pox and rubella affect 100% of exposed individuals; however, peanut allergy affects only 1% of children and the risk to other people of that immunisation cannot be justified. Researchers who are working on immunisations need to devise a secondary immunisation strategy where people who already have the allergy are treated.

4.10   Conclusions

In conclusion, peanut allergy is probably a life-long danger in most cases. The only way to find out if peanut allergy persists, or has resolved, is to do challenges, and the number of centres doing challenges needs to increase in order to be able to remove this diagnosis from people for whom it is no longer justified. If the allergy is severe, it usually stays severe, but if it is mild, but persists, it may worsen. Exposure to peanut allergen is likely, and unfortunately at present there is no cure.

Doctors and providers of food information and community support to allergy sufferers must encourage them to be proactive when they are eating away from home or consuming foods prepared by others. They must be trained to self-treat with epinephrine because at the moment there is nothing else that can help.

4.11   References

1.   Hourihane J.O'B., Roberts S.A., Warner J.O. Resolution of peanut allergy: case-control study. *British Medical Journal*, 1998, 316 (7140), 1271-5.

2.   Skolnick H.S., Conover-Walker M.K., Koerner C.B., Sampson H.A., Burks W., Wood R.A. The natural history of peanut allergy. *Journal of Allergy and Clinical Immunology*, 2001, 107 (2), 367-74.

3.   Vander Leek T.K., Liu A.H., Stefanski K., Blacker B., Bock S.A. The natural history of peanut allergy in young children and its association with serum peanut-specific IgE. *Journal of Paediatrics*, 2000, 137 (6), 749-55.

4.   Wuthrich B., Dascher M., Borelli S. Kiss-induced allergy to peanut. *Allergy*, 2001, 56, 9, 913.

5.   Li X.M., Schofield B.H., Huang C.K., Kleiner G.I., Sampson H.A. A murine model of IgE-mediated cow's milk hypersensitivity. *Journal of Allergy and Clinical Immunology*, 1999, 103 (2 Pt 1), 206-14.

6.   Li X.M., Kleiner G., Huang C.K., Lee S.Y., Schofield B., Soter N.A., *et al.* Murine model of atopic dermatitis associated with food hypersensitivity. *Journal of Allergy and Clinical Immunology*, 2001, 107 (4), 693-702.

7.   Grundy J., Bateman B., Gant C., Matthews S., Dean T., Arshad H. Peanut allergy in three year old children - a population based study. *Journal of Allergy and Clinical Immunology*, 2001, 107 (2), S231.

8.   Hourihane J. O'B., Kilburn S.A., Dean T.P., Warner J.O. Clinical characteristics of peanut allergy. *Clinical and Experimental Allergy*, 1997, 27 (6), 634-9.

9.   Sicherer S.H., Furlong T.J., Munoz-Furlong A., Burks A.W., Sampson H.A. A voluntary registry for peanut and tree nut allergy: characteristics of the first 5149 registrants. *Journal of Allergy and Clinical Immunology*, 2001, 108 (1), 128-32.

10.   Ewan P.W., Clark A.T. Long-term prospective observational study of patients with peanut and nut allergy after participation in a management plan. *Lancet*, 2001, 13, 357 (9250), 111-5.

11.   Bock S., Munoz-Furlong A., Sampson H. Fatalities due to anaphylactic reactions to foods. *Journal of Allergy and Clinical Immunology*, 2001, 107 (1), 191-3.

12.   Niemann L.M., Hlywka J.J., Hefle S.L. Immunochemical analysis of retail foods labelled as "may contain peanut" or other similar declaration: implications for food allergic individuals. *Journal of Allergy and Clinical Immunology*, 2000, 105(1), S188.

13.   McCabe M., Lyons R.A., Hodgson P., Griffiths G., Jones R. Management of peanut allergy, *Lancet*, 2001, 12, 357 (9267), 1531-2.

14.   Beyer K., Morrow E., Li X.-M., Bardina L., Bannon G.A., Burks A.W., Sampson H.A. Effects of cooking methods on peanut allergenicity. *Journal of Allergy and Clinical Immunology*, 2001, 107 (6), 1077-81.

15.   Maleki S.J., Chung S.-Y., Champagne E.T., Raufman J.-P. The effects of roasting on the allergenic properties of peanut proteins. *Journal of Allergy and Clinical Immunology*, 2000, 106 (4), 763-68.

5. ALLERGENICITY OF SOYA BEAN LECITHINS AND IDENTIFICATION OF A POLLEN-RELATED ALLERGEN IN A COMMERCIAL SOYA PROTEIN ISOLATE

Stefan Vieths and Jörg Kleine-Tebbe

5.1   Introduction

The first part of this paper covers the study conducted at the Paul-Ehrlich-Institut, together with a company from the food industry, on allergenicity of soya bean lecithins (1). The second part covers a study also conducted recently at the Paul-Ehrlich-Institut in collaboration with the University Hospital of Leipzig, in which a pollen-related allergen was identified in soya protein isolate that was used in a special product on the German market, and caused a surprisingly high rate of severe reactions (2).

5.2   Prevalence

It is quite clear that soya bean allergy occurs in infants as well as in adults; however, it is generally accepted that it is less severe and less frequent than peanut allergy. The epidemiological data on soya bean allergy are very poor, and there are inconsistent data on the identity of soya bean allergens. There are very few population-based studies on food intolerance; one study has been conducted by Dr Moneret-Vautrin (3), two in the Netherlands, by Brugman *et al.* (4), and by Niestijl Jansen *et al.* (5), and one in the UK by Young *et al.* (6), who found that, from 16 food studies investigated, soya bean was the least frequent allergenic food, with a prevalence of 0.3%.

5.2.1   In children

The frequency of soya bean allergy varies from 1.2% in a study conducted by Rance *et al.* (7), up to 28% in a special population of food allergic patients with atopic dermatitis, studied by double-blind placebo-controlled food challenges by Sampson and Ho (8). There is clearly a large variation in the reported frequencies of soya bean allergy.

5.2.2 In adults

A recent study on the prevalence of allergy in adults by Etesamifar and Wüthrich (9) found that, in 383 fruit-allergic patients, 9% were allergic to a soya bean according to history and skin prick tests or serum IgE levels. The study by Mistereck *et al.* (10) also reported a prevalence of allergy of 9% in 250 patients with suspected food allergy. André *et al.*'s study (11) found that, in 60 patients with food anaphylaxis, 3.3% (2 cases) were shown to be caused by soya bean.

In the study by Foucard & Yman in Sweden (12), 61 cases of severe food reactions were investigated, and it was discovered that 33% were caused by peanut, and 26% by soya bean; this is a very high rate of severe reactions to soya bean. It was also found that four fatal reactions to soya were caused by soya protein hidden in meat products containing a total of between 2.2 and 7% of soya protein. The ingested amount corresponded to between 1 and 10 g of soya, and the patients who died were aged between 10 and 17 years and had no previously known allergy to soya. However, this study has been criticised because it is possible that there could have been peanut protein in the samples tested; the samples were then analysed for peanut protein, but the assay used was not very sensitive, and was able to detect only up to 50 ppm peanut protein. It is important to bear in mind that 100 μg peanut protein may cause an allergic reaction, so it is best to use an assay with a sensitivity to 1-2 ppm protein. The retrospective analysis was thus not fully convincing in determining whether peanut protein was present; therefore, it could not be confirmed that the deaths were definitely a result of allergic reaction to soya protein.

5.3   Threshold Levels

There are very few data on threshold levels in soya bean allergy. Bock (13) reported 1 g soya bean in dry matter, which is more than a hundred-fold greater than the threshold level for peanut allergy reported by Hourihane *et al.* (14).

5.4   Soya Protein Allergens

The official allergen list of the Allergen Nomenclature Subcommittee of the WHO International Union of Immunological Societies lists three soya allergens. Two of them, Gly m 1 (hydrophobic soya bean protein) and Gly m 2 (disease response protein), are soya bean hull proteins involved in the asthma outbreaks in Barcelona in 1987 and 1988 (15). Gly m 3, discovered by Rihs *et al.* (16) is a profilin. It was found that there was a positive IgE response to profilin in 9 out of 13 patients with atopic dermatitis and suspected food allergies; however, it was not confirmed that these 9 patients were indeed allergic to soya beans. The Kunitz-trypsin-inhibitor has been identified as an important allergen in patients with bakers' asthma (17); however, it is a respiratory allergen rather than a food allergen.

There has been extensive research on allergenic soya bean proteins by Ogawa *et al.* in Japan (18), who identified Gly m Bd 28k, a vicilin-like protein, Gly m Bd 30k, a cysteine proteinase, and Gly m Bd 60k, an α-conglycinin, as soya bean allergens. However, soya-bean-sensitive patients with atopic dermatitis were studied, and sensitivity was concluded from IgE binding and not from challenge experiments; therefore, the evidence is not conclusive that these three proteins are indeed soya bean allergens.

There are also studies that report that subunits of glycinin were allergenic in soya beans (19-21). Glycinin consists of two subunits: both glycinin G1 and glycinin G2 are comprised of acidic polypeptides of between 30 and 45 kDa and basic polypeptides of between 18 and 20 kDa. Using the serum of seven individuals with a clear history of soya bean allergy, Zeece *et al.* (20) identified the acidic subunit of glycinin G1 as an allergen, and Helm *et al.* (21) identified the basic subunit of glycinin G2 as an allergen in soya beans using serum from seven patients with a positive double-blind placebo-controlled challenge to soya bean.

5.5   Lecithin Study

5.5.1   Lecithin samples

Six well-characterised lecithin samples were obtained from "The Lecithin People" in Hamburg, who were interested in the residual allergenic potency of their products (1). The manufacturing procedure for lecithin is given in Fig. 5.1: the crude lecithin is selected, blended and adjusted to obtain standard lecithins. Of the two food-grade lecithins used in the study, one had 3,100 ppm residual protein, the other had 120 ppm residual protein. Other lecithins were obtained by further fractionation: ethanol extraction resulted in a phosphatidylcholine-depleted fraction with 13 ppm of residual protein, and enriched fractions that contained less than 20 ppb of protein. The de-oiled lecithin had 65 ppm of residual protein.

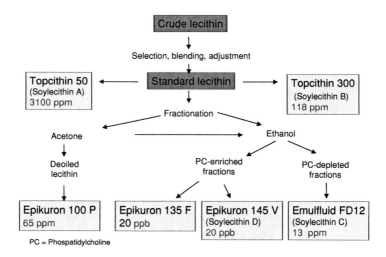

Fig. 5.1. Production of soya bean lecithins investigated for residual allergenic potency

The two lecithins with the highest amount of protein were the food-grade lecithins. The de-oiled lecithins are mainly used for dietary purposes, and the phosphatidylcholine-enriched fractions are used for medical applications. The phosphatidylcholine-depleted product is a special emulsifier.

5.5.2 IgE reactivity - methodology

Sera from six patients with a clear history of allergy to soya bean were selected from the sera collection at the Paul-Ehrlich-Institut. The patients' reactions to soya beans are shown in Table 5.I, and ranged from oral allergy syndrome to gastrointestinal symptoms, or dyspnoea and nausea. All patients were also allergic to pollen. An extract was prepared from raw soya beans and from cooked soya beans and IgE reactivity as determined by an enzyme allergosorbent test were used to compare the allergenicity of the two forms of the protein. Allergen extracts were also prepared from all the soya bean lecithins.

TABLE 5.I
Symptoms of patients in response to soya bean protein

Soya-bean-allergic Patients

Patient	EAST class soya bean		Symptoms	Pollen allergy
	raw	cooked		
1	3	2	dyspnoea, nausea	yes
2	4	4	oral allergy syndrome (OAS)	yes
3	3	3	OAS, gastrointestinal	yes
4	2	0	gastrointestinal	yes
5	2	0	OAS, urticaria	yes
6	2	0	OAS to sprouts	yes

5.5.2.1 Results

All six patients reacted to the raw soya bean, whereas only three showed specific IgE to an extract from cooked soya beans. There was no IgE binding of extracts of soya bean lecithins with the sera that had shown specific IgE antibody to the raw soya bean extract. There was also no IgE binding to the two lecithins containing less than 20 ppb of soya protein, but sera from patients 1, 2 and 3 - those who reacted to the cooked soya bean extract - showed enzyme allergosorbent test (EAST) classes of 1-4 to the four other lecithins.

5.5.3   Immunoblot experiment

An immunoblot experiment was conducted with an extract from raw soya bean and from cooked soya beans, using the serum of patients 1, 2, and 3, which showed reactivity to the cooked soya bean. One band was present at about 14 kD, together with another high molecular weight band using the raw soya bean extract. The 14 kD band was still present with the cooked soya bean protein, whereas most of the other bands that were visible with the raw soya bean extract had gone.

The immunoblot test also used a monoclonal antibody raised against the acidic subunit of glycinin (22), and this antibody recognised the same protein band as the patients' serum for both raw and cooked soya protein.

5.5.4   Cross-linking of cell-bound IgE

The ability of proteins in the lecithins to cross-link cell-bound IgE was investigated using an animal model.

5.5.4.1 Animal model

BALB/c mice were immunised with soya bean extract to produce a specific IgE response, and murine sera for passive sensitisation of rat basophil leukaemia cells sublime (RBL-2H3). Murine antibodies were bound to high affinity receptors on these permanently cultivated cells and a dose-dependent challenge was performed with allergen; the released mediators were measured using β-hexosaminidase - an enzyme that is present in the same granula as histamine. This assay is a valid replacement for passive cutaneous anaphylaxis, and is replacing a lot of animal experiments.

5.5.4.2 Results

At the highest dose given, all lecithins that contained residual protein (>20 ppb) were able to induce allergen-specific mediator release in this biological *in vitro* model.

5.5.4.3 Conclusions

All lecithins that contained protein traces in the ppm range bound soya bean-specific IgE from allergic patients with IgE to heated soya beans. It is most likely that the acidic subunit of glycinin is the heat-stable allergen in the soya bean lecithins recognised by IgE from the soy bean-allergic participants of our study.

An important question is whether or not the quantities of soya proteins that are introduced into commercial food products, via the use of these lecithins, are able to elicit allergic reactions in the consumer.

If the lecithin with the highest amount of residual proteins at 0.3 g/100 g of lecithin is used as an example, and an application level in foods of approximately 1% is assumed, then 100 g of chocolate may contain 3 mg of soya protein by carry-over in lecithins. At the moment it is not clear whether or not 3 mg of soya protein (if one person eats 100 g chocolate, then the dose of soya protein would be about 3 mg) are able to elicit an allergic reaction.

If this 3 mg level is compared with the threshold level reported by Bock *et al.* (6), it is a 100-fold below the reported threshold level for soya protein allergenicity. However, the possibility that highly sensitive patients may react to this cannot be discounted.

5.6   Bet v1-related Allergen in Soya Bean

A second project being conducted by the Paul-Ehrlich-Institut is on severe reactions to a dietary product containing a soya protein isolate (2,23). The product, Almased, consists of 25% granulated yoghurt, 25% granulated honey, 50% commercial soya protein isolate, and a vitamin complex. A single serving is 2 tablespoons, which are rehydrated with water, milk, or juice.

The product has been on the market since 1984; however, since 1998, there has been an increasing number of reports of allergic reactions to the product, and several consumers complained to the Company that they had had severe allergic reactions after eating the product. More than 100,000 packages of Almased are sold in Germany per month.

The soya protein isolate that the product contains is prepared by a simple procedure from defatted soya bean flakes. The protein is extracted at a pH of 8.0 to 8.5, in slightly basic conditions. Then it is neutralised and the product is complete after spray drying.

Kleine-Tebbe studied the first patient who complained about this product, and contacted all of the people who complained about reactions to the product. Data were collected on 20 patients, and their sera were collected (2).

Almost all of the patients studied had suffered a reaction after ingesting this dietary product for the first time and they were not aware that they were allergic to soya beans. The allergic reaction in most of the patients started with oral allergy syndrome, or itching of the eyes, then erythema, and many of them had more severe reactions, including face swelling, throat tightness, and dyspnoea. Urticaria, cardiovascular or gastrointestinal problems, vomiting and ear swelling also occurred.

Sixteen of the 20 patients were highly allergic to birch and had very high IgE titres to Bet v 1. Most of the patients were also allergic to other pollens and many of them – 12 at least - reported pollen-related allergy to typical birch pollen related food, such as apple, hazelnut, peach or carrot, or other fruits.

The first step in identifying the allergens in the product was to exclude patients who were sensitised to milk, or perhaps to pollen included in the honey, or to components of the vitamin mixture. To do this, allergen extracts were prepared from all the constituents of the product, and were used for skin prick tests. Patients were also subjected to an EAST for IgE to these components. None of the patients tested was sensitive to yoghurt, honey or the vitamins.

The IgE determinations showed that most patients had a very high IgE response to birch, and moderate IgE levels to soya beans in the commercial assay.

5.6.1 Allergen identification

The allergen was identified by means of blotting studies using an allergen extract prepared from the soya protein isolate. A paper by Crowell *et al.* (24) contains the results of a study on the messenger RNA of protein that was inducible in soya bean roots and in primary leaves of soya bean by various kinds of stress, including fungal proteins and pesticides. For example, Crowell *et al.* demonstrated that the herbicide Paraquat induced messenger RNA synthesis. This RNA is called SAM 22, and 48% of the deduced amino acid sequence is identical to the major birch pollen allergen Bet v 1. A pBluescript plasmid containing the DNA of the SAM 22 protein was provided by Crowell *et al.* (24), and a PCR was performed to add restriction sites; SAM 22 was cloned into an expression vector, and then the protein was expressed in *E. coli* BL21, and the non-fusion protein was purified by chromato-focusing. This protein was used in addition to other protein extracts to study the patient group.

Sixteen of the patients – all 16 who were allergic to birch – presented IgE antibody to recombinant SAM 22. In the immunoblot of the soya protein isolate, two out of the 20 showed IgE binding to a 35-38 kD band, which probably represents a glycinin subunit, and one patient showed IgE binding to a 22 kD band.

Inhibition experiments were conducted using an extract of the soya protein isolate. Positive control of the soya isolate inhibits 100% of the IgE binding to the soya isolate itself and, the recombinant SAM 22 also inhibits almost all IgE reactivity to the soya protein isolate. It is clear that almost all IgE binding capacity of the soya protein isolate is related to birch; therefore, it can be concluded that there is a Bet v 1 related allergen in soya bean.

5.7   Conclusions

It has been shown that food lecithins contain proteins, which bind IgE from soya-bean-allergic patients. The soya protein levels in processed food may be a few milligrams per 100 g of final products; however, soya protein sensitivity is thought to be a 100-fold higher than this. However, allergic reactions in highly sensitised patients who love to eat lots of chocolate cannot be completely discounted.

The results of the study on soya bean lecithins are not in accordance with those of Awazuhura *et al.* (19), who reported very low IgE binding activity in commercial soya bean lecithins. Perhaps the selection criteria of patients was different from that in Müller's study (1).

The PR10 pathogenesis-related protein, SAM 22, appears to be responsible for severe reactions to soya protein isolate in patients sensitised to the major birch allergen Bet v 1. The main reason that this allergen has been overlooked in previous studies may be that this soya protein isolate is consumed more or less in a native state. It is not heat-treated and perhaps the heat-labile Bet v 1 allergen is de-activated in almost all soya bean products, except this one, and perhaps soya bean sprouts.

A risk assessment is required for other soya bean protein containing products - for example, by double-blind challenges in patients with birch-pollen-related food allergy but without known allergy to soya beans. So far, the abundance of this SAM 22 protein in other soya bean protein products is unknown.

5.8   References

1.   Müller U., Weber W., Hoffmann A., Franke S., Lange R., Vieths S. Commercial soybean lecithins: a source of hidden allergens? *Zeitschrift für Lebensmittel Untersuchung und –Forschung*, 1998, 207 (5), 341-51.

2.   Kleine-Tebbe J., Vieths S., Franke S., Jahreis A., Rytter M., Haustein U.-F. Severe allergic reactions to a dietary product containing soy protein due to IgE-mediated cross-reactivity and hypersensitivity to Bet v1. *Allergo Journal*, 2001, 10, 154-9.

3.   Kanny G., Moneret-Vautrin D.A., Flabbee J., Beaudouin E., Morisset M., Thevenin F. Population study of food allergy in France. *Journal of Allergy and Clinical Immunology*, 2001, 108 (1), 133-40.

4.   Brugman E., Meulmeester J.F., Spee-van der Wekke A., Beuker R.J., Radder J.J., Verloove-Vanhorick S.P. Prevalence of self-reported food hypersensitivity among school children in The Netherlands. *European Journal of Clinical Nutrition*, 1998, 52 (8), 577-81.

5.   Niestijl Jansen J.J., Kardinaal A.F.M., Huijbers G., Flieg-Boerstra B.J., Martens B.P.M., Ockhuizen T. Prevalence of food allergy and intolerance in the Dutch population. *Journal of Allergy and Clinical Immunology*, 1994, 93, 446-56.

6.   Young E., Stoneham M.D., Petruckevitch A., Barton J., Rona R. A population study of food intolerance. *Lancet*, 1994, 343 (8906), 1127-30.

7.   Rance F., Kanny G., Dutau G., Moneret-Vautrin D.A. Food hypersensitivity in children: Clinical aspects and distribution of allergens. *Pediatr. Allergy Immunol.*, 1999, 10, 33-8.

8.   Sampson H.A., Ho D.G. Relationship between food-specific IgE concentrations and the risk of positive food challenges in children and adolescents. *Journal of Allergy and Clinical Immunology*, 1997, 100, 444-51.

9.   Etesamifar M., Wüthrich B. IgE-vermittelte Nahrungsmittelallergie bei 383 Patienten unter Berücksichtigung des oralen Allergie-Syndroms. *Allergologie*, 1998, 21, 451-7.

10.   Mistereck A., Lange C.E., Sennekamp J. Soya – a frequent food allergen. *Allergologie*, 1992, 15, 304-5.

11.   André F., André C., Colin L., Cacaraci F., Cavagna S. Role of new allergens consumption in the increased incidence of food sensitization in France. *Toxicology*, 1994, 92, 77-83.

12.   Foucard T., Malmheden Yman I. A study on severe food reactions in Sweden - is soy protein an underestimated cause of food anaphylaxis? *Allergy*, 1999, 54, 261-5.

13.   Bock S.A., Lee W.Y., Remigio L.K., May C.D. Studies of hypersensitivity reactions to foods in infants and children. *Journal of Allergy and Clinical Immunology*, 1978, 62 (6), 327-34.

14. Hourihane J., Kilburn S., Nordlee J., Hefle S., Taylor S., Warner J. An evaluation of the sensitivity of subjects with peanut allergy to very low doses of peanut protein: A randomized, double-blind placebo-controlled food challenge study. *J. Allergy Clin. Immunol.*, 1997, 100, 596-600.

15. Gonzalez R., Polo F., Zapatero L., Caravaca F., Carreira J. Purification and characterization of major inhalant allergens from soybean hulls. *Clinical and Experimental Allergy*, 1992, 22 (8), 748-55.

16. Rihs H.P., Chen Z., Rueff F., Petersen A., Rozynek P., Heimann H., Baur X. IgE binding of the recombinant allergen soybean profilin (rGly m 3) is mediated by conformational epitopes. *Journal of Allergy and Clinical Immunology*, 1999, 104 (6), 1293-301.

17. Baur X., Pau M., Czuppon A., Fruhmann G. Characterization of soybean allergens causing sensitization of occupationally exposed bakers. *Allergy*, 1996, 51 (5), 326-30.

18. Ogawa A., Samoto M., Takahashi K. Soybean allergens and hypoallergenic soybean products. *J Nutr Sci Vitaminol*, 2000, 46 (6), 271-9.

19. Beardslee T.A., Zeece M.G., Sarath G., Markwell P. Soybean glycinin G1 acidic chain shares IgE epitopes with peanut allergen Ara h3. *International Archives of Allergy and Immunology*, 2000, 123, 299-307.

20. Zeece M.G., Beardslee T.A., Markwell J.P., Sarath G. Identification of an IgE-binding region in soybean acidic glycinin G1. *Food Agric Immunology*, 1999, 11, 83-90.

21. Helm R.M., Cockrell G., Connaughton C., Sampson H.A., Bannon G.A., Beilinson V., Nielsen N.C., Burks A.W. A soybean G2 glycinin allergen. 2. Epitope mapping and three-dimensional modeling. *International Archives of Allergy and Immunology*, 2000, 123 (3), 213-9.

22. Carter J.M., Lee H.A., Mills E.N.C., Lambert N., Chan H.W.-S., Morgan M.R.A. Characterisation of polyclonal and monoclonal antibodies against glycinin, (11S storage protein) from soya (*Glycine max*). *J. Sci. Food Agric.*, 1992, 58, 75-82.

23. Scherf H.P., Bauer P. Allergic reaction to a dietary powder containing soy protein in a pollinosis patient. *Allergologie*, 2001, 23, 190-4.

24. Crowell D.N., John M.E., Russell D., Amasino R.M. Characterization of a stress-induced developmentally regulated gene family from soybean. *Plant Molecular Biology*, 1992, 18, 459-66.

25. Awazuhara H., Kawai H., Maruchi N. Major allergens in soybean and clinical significance of IgG4 antibodies investigated by IgE- and IgG4-immunoblotting with sera from soybean-sensitive patients. *Clinical and Experimental Allergy*, 1997, 27 (3), 325-32.

6. SEAFOOD ALLERGY AND ALLERGENS

Cristina Pascual, M.T. Belver, A. Valls, and M. Martin-Esteban

6.1   Introduction

This paper covers allergic reactions to seafood, i.e. those that have an immunological mechanism.

Adverse reactions can be IgE-mediated or non-IgE-mediated. There are tools for defining and diagnosing the IgE-mediated reactions; less is known about non-IgE-mediated reactions. This paper covers mainly IgE-mediated reactions to seafood.

6.2   Prevalence

Seafood is an important food in the staple diet in the northern countries in Europe; Spain and Portugal have the same level of fish intake as Sweden and Norway, for example.

Table 6.I shows the prevalence of food allergy in three groups of patients; two of the groups were adult patients and one group was children. The two groups that comprised 417 children and 51 adults, respectively, came from Madrid, and the group of 142 adults came from a subtropical area in the Canary Islands. All of the groups were matched in income and social status and the two adult groups were matched for age. The pattern of food allergy suggests that the staple diet in the tropical area of the Canary Islands is different from that of the other groups, and this difference in staple diet is significant in the difference in prevalence and pattern of allergy in each of the groups.

Fish allergy has been recognised for some time because, in 1937 De Besche from Denmark published a paper explaining that fish was a complete allergen - an inhalant allergen, a contact allergen and also an alimentary allergen. Fish allergy is rare in the United States; however, there are many cases of fish allergy in Spain.

TABLE 6.I

Incidence of food allergy in Spain (allergy outpatient setting)

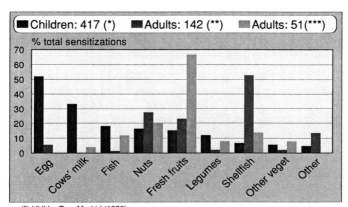

(*) HUI La Paz, Madrid (1996)
(**) HU NS del Pino, Las Palmas GC (1988-94)
(***) HU 12 Octubre, Madrid (1995)

6.3 Allergen in Fish

A paper by Tuft (3) was published relating to a high cross-reactivity between the species with IgG antibody. Nothing was known about IgE at that time. In the late 60s, or early 70s Aas (4) published a large study about cod allergen, which was first called M Allergen, and was a paradigm of how a food allergen should be studied. In 1975 Elsayed (5) sequenced the allergen and it was discovered that parvalbumin was the major allergen of fish. The cross-reactivity was established between many species of fish.

6.4 Allergenicity of Different Species of Fish

The allergenicity of six species of fish was studied in children. Flat fish were among those chosen for use in the study because, in Spain, flat fish are the first fish introduced into children's diets. At 2 years of age, children are traditionally given albacore, which is very similar to tuna fish. The species of fish used in this study were meagrin, sole and the whitch flounder. Other non-flat fish tested were cod, albacore, and hake, because these species have been cited as being allergenic in many studies (4,6)

The first study on the incidence of fish allergy was conducted in the late 1980s (7). Eighteen per cent of the children examined in the study who had food allergy were allergic to fish. Another study was conducted 5 years later (2), and the incidence of allergy to fish increased. A preventive measure was introduced in that introduction of fish into the diet of children at high risk of developing allergy to fish was postponed in the first year of life, so fish allergy was lower in the first year of life, but appeared in the second year because the effect of the preventive measure was to postpone the incidence of allergy by one year.

6.5   Symptoms of Fish Allergy

The symptoms of fish allergy were investigated in 79 children. Most cases suffered acute cutaneous symptoms - mostly urticaria and angioedema, and some suffered oral allergy syndrome (OAS) and worsening of atopic dermatitis. Vomiting and diarrhoea were also experienced. It is important to define what constitutes anaphylaxis. In Europe, systemic anaphylaxis involves the patient suffering from shock; however, in the USA, anaphylaxis is defined as a reaction involving two or more organs. Therefore, if there is generalised urticaria and vomiting, this is classified as an anaphylactic reaction in the USA.

A cutaneous prick test was used to diagnose fish allergy in this study, together with *in vitro* IgE and challenge testing. If the patient was under 2 years of age, an open challenge was given; however, if the patient was over 2 years old, a single-blind placebo-controlled challenge was performed.

All 79 children had positive skin prick tests to the six species; however, only 20% of children had positive IgE to albacore. It was not possible to determine whether one species of fish was more significant than another in eliciting an IgE response; however, following challenge testing, 39% of children tolerated tuna fish, even if they had a positive IgE response to tuna.

6.6   Parvalbumin Protein

A Japanese research group discovered that the parvalbumin protein was not the most allergenic; there was another protein that was possibly responsible for sensitisation to fish.

Reinforcing the results demonstrated in the study by Pascual *et al.* (9), where positive IgE responses to tuna fish were demonstrated, but children showed tolerance to oral challenge with tuna fish, it has been shown that the scombriforme family – yellow fin tuna, albacore, and also swordfish, may be less allergenic than other species of fish. Indeed, in Spain, parents of children at high risk of fish allergy are now encouraged to provide tuna or swordfish as the first fish in the diet, which is the way that fish is introduced into children's diets in the United States. Children only eat tuna fish; canned tuna fish has a higher level of protein denaturisation than fresh tuna.

Atopic dermatitis patients were shown by Larramendi *et al.* (10) to have worse dermatitis after ingestion of fish. These patients were put on a diet, and were challenged with fish after 2 or 3 years in order to see if they could tolerate fish. Patients developed urticaria and angioedema in response to fish intake, so it was decided that, when some consumption of some fish, such as tuna or albacore, was tolerated, patients were advised to keep eating that fish.

Patients were followed up for several years. In the first study (11), the mean age of patients was 1.6 year, but after approximately 2 years the patients showed symptoms through inhalation and other symptoms through contact. All these patients were already on a fish-free diet. However, the symptoms caused through inhalation of steam from cooked fish again caused urticaria and angioedema. These symptoms occurred in patients that were more sensitive to fish. The same symptoms occurred as a result of cutaneous contact. These symptoms all occurred in children who reported symptoms when fish was boiled, fried, or just handled at home. Two children suffered allergic symptoms after visiting a fish market; air samples were collected from the fish market and it was shown that small amounts of fish allergen were present in the air of this environment.

Research was also conducted on the allergenicity of salmon. Salmon is imported from Norway to Spain and is relatively cheap, so is becoming more popular. It was thought that a difference in allergenicity between salmon and other species could be identified. Salmon was boiled and compared with other allergenic fish. Results were published in 1996. Immunoblotting of raw salmon, boiled salmon, and the vapour from boiled salmon were examined. It was discovered that there were allergens in the vapour that could be obtained from boiling fish at home. It is important to be aware of this if a member of the household is allergic to fish, as repeated exposure to fumes from cooking fish can enhance sensitivity.

The children in this study were followed up for more than 16 years. Prevalence of allergy to cows' milk in children drops after 2 years, and, after 6 years of age, a very small number of children retained their allergy to cows' milk, especially those that were sensitised to casein, because casein is considered as pernicious in cows' milk allergy. After more than 10 years, 80% of the children were allergic to fish and retained their allergy.

6.7   Shellfish and Mollusc Allergy

Table 6.I also shows incidence of shellfish allergy; 5% of children were allergic to shellfish, between 12 and 13% of adults in Madrid demonstrated fish allergy, and 55% of adults with food allergy in the Canary Islands were allergic to shellfish (12). The intake of limpets is very high in this area so most of these cases are due to mollusc consumption; it is unusual to find references to allergy to molluscs in the literature. The Northern Brown Shrimp is the species used to identify IgE against shrimp species. The major allergen of oyster has also been identified; however, all of the shellfish have a very similar reactivity. The allergenic proteins are mostly tropomyosins of between 34 and 36 kD and the cross-reactivity is between 80 and 90% between all crustacea, and is slightly lower for molluscs.

Allergy to the German cockroach was studied in a population of children studied by Crespo *et al.* (13) that were allergic to shrimp. There was significant cross-reactivity due to tropomyosin. Immunoblotting was also performed; the most important band was tropomyosin, and the inhibition was in that area.

There have been papers published indicating that conducting immunotherapy with dermatophagoides in some patients can enhance food allergy to shrimp. It is not something that is often proved; however, tropomyosin is a panallergen found in many species.

6.8   Allergenic Contaminants of Fish

Anisakiasis is a parasitic disease, and anisakis is a nematode. Consumption of raw or undercooked fish containing these nematodes can result in parasitic disease. The larva can be destroyed by freezing, and, in Holland, a law was passed in 1968 that fish should not be consumed without prior freezing. In Spain, some fish is habitually eaten raw, e.g. anchovies, and this provides a human model of IgE response in cases where parasitic disease has been contracted. Anisakis can be considered as a pseudoallergen because it is a contaminant of food, as is dermatophagoides, which can be present in the flour.

A paper published in Japan in 1990 reported that people were suffering urticaria and angioedema after ingestion of well-cooked fish. This was a new development for anisakiasis, because it was associated as being a problem linked to consumption of raw fish. The life cycle of anisakis is shown in Fig. 6.1, and shows that the nematodes cannot reproduce in humans. The survival rate in humans is 3 weeks, and the larvae migrate into the stomach mucosa or intestinal mucosa, and can cause gastroallergic anisakiasis, which is characterised by urticaria and angioedema and vomiting. Anaphylactic shock as a result of ingestion of the contaminated fish is also possible.

Sixty-three per cent of the Japanese population have IgG antibody against anisakis, but nothing is known about sensitivity to anisakis in Europe.

The level of sensitisation to anisakis in Europe was determined by Pascual et al. (14). Several groups were used to determine sensitivity to anasikis – one group with paediatric shell fish allergy, one with paediatric fish allergy, one with urticaria/oeosinophillia to fish – this group comprised adults, and a paediatric group with high total IgE, who were asthmatic. The proportion of the groups who demonstrated positive IgE response to anisakis are shown in Fig. 6.2.

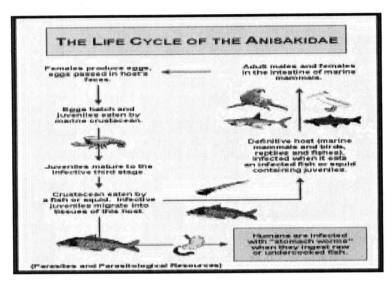

Fig. 6.1. The life cycle of the Anisakidae

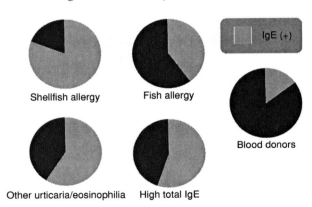

Fig. 6.2. Specific IgE to Anisakis simplex

Nematodes have a secretory protein that allows them to break the mucosal wall and enter the stomach, or the intestine. This is a proteolytic protein of between 14 and 17 kD. The importance of this allergen in sensitisation was investigated in live nematodes (15). Anisakis larvae were cultivated and extracted from cod. The same groups as shown in Fig. 6.2 were used to determine allergenicity of the secretory protein. Sensitivity to the allergen was shown to be high.

A challenge test was performed using dead encapsulated larvae in those patients who had urticaria and angioedema following contact with raw fish, and the challenge was negative. It was concluded that, in order to have a reaction, the patient must have contact with a live worm.

6.9   References

1.   De Besche A. On asthma bronchiale in man provoked by cat, dog, and different other animals. *Acta. Med. Scand.*, 1937, 42, 237-55.

2.   Crespo J.F., Pascual C.Y., Burks A.W., Helm R.M., Martin Esteban M. Frequency of food allergy in a pediatric population from Spain. *Pediatric Allergy and Immunology.*, 1995, 6, 39-43.

3.   Tuft L., Blumstein G.I. Studies in food allergy. V. Antigenic relationship among members of fish family. *Journal of Allergy*, 1946, 17, 329-39.

4.   Aas K. Studies of hypersensitivity to fish: clinical study. *International Archives of Allergy*, 1966, 29, 346-63.

5.   Elsayed S., Bennich H. The primary structure of allergen M from cod. *Scandinavian Journal of Immunology*, 1975, 4, 203-8.

6.   Pascual C., Martin Esteban M., Crespo J.F. Fish allergy: Evaluation of the importance of cross-reactivity. *Journal of Pediatrics*, 1992, 121, 29-34.

7.   Pascual C.Y., Larramendi C.H., Martin Esteban M., Fiandor A., Ojeda J.A. Fish allergy and fish allergens. *Journal of Allergy and Clinical Immunology*, 1988, 81, 264.

8.   Sakaguchi M., Toda M., Ebihara T., Irie S., Hori H., Imai A., Yanagida M., Miyazawa H., Ohsuna H., Ikezawa Z., Inouye S. IgE antibody to fish gelatin

(type I collagen) in patients with fish allergy. *Journal of Allergy and Clinical Immunology*, 2000, 106 (3), 579-84.

9. Pascual C.Y., Crespo J.F., Sanchez-Pastor S., *et al.* The Importance of Fish in IgE - Mediated Food-Hypersensitivity. *ACI News*, 1995, 7, 73-5.

10. Larramendi C.H., Martin Esteban M., Pascual Marcos C., Fiandor A., Diaz Pena J.M. Possible consequences of the elimination diet in asymptomatic immediate fish hypersensitivity. *Allergy*, 1992, 47, 490-4.

11. Crespo J.F., Pascual C., Dominguez C., Ojeda I., Martin F., Martin Esteban M. Allergic reactions associated with airborne fish particles in IgE-mediated fish hypersensitivity patients. *Allergy*, 1995, 50, 257-61.

12. Castillo R., Carrillo T., Blanco C., Quiralte J., Cuevas M. Shellfish hypersensitivity: clinical and immunological characteristics. *Allergol. Immunopathol.*, 1994, 22, 83-7.

13. Crespo J.F., Pascual C., Helm R., Sanchez-Pastor S., Ojeda I., Romualdo L., Martin Esteban M.M., Ojeda J.A. Cross-reactivity of IgE-binding components between boiled Atlantic shrimp and German cockroach. *Allergy*, 1995, 50, 918-24.

14. Pascual C., Crespo J.F., Ortega N., Ornia N., San-Martin M.S., Martin Esteban M. High prevalence of sensitization to Anisakis simplex in patients with increased levels of total IgE. *Journal of Allergy and Clinical Immunology*, 1996, 97, 233.

15. Arrieta I., del Barrio M., Vidarte L., del Pozo V., Pastor C., Gonzalez-Cabrero J., Cardaba B., Rojo M., Minguez A., Cortegano I., Gallardo S., Aceituno E., Palomino P., Vivanco F., Lahoz C. Molecular cloning and characterisation of an IgE-reactive protein from Anisakis simplex: Ani s 1. *Mol. Biochem. Parasitol.*, 2000, 107 (2), 263-8.

7. SESAME ALLERGY

Dr Denise-Anne Moneret-Vautrin

7.1   Introduction

There is a well-known correlation between the level of rice consumption in Japan and the prevalence of allergy in that country. It is the same case with sesame. China and Japan are the main global producers of sesame, and the prevalence of allergy to sesame is higher in these countries (1).

Sesame seeds are imported into European countries; 2,400 tons of sesame seeds were imported into France in 2001 and 28,000 tons into the Netherlands. France imported 400 tons of sesame oil. These disparities between the levels of imports in different countries are reflected in the differing incidence of sesame allergy in various countries.

Sesame seeds belong to Dicotyledones in the *Pedaliaceae* family, genus: sesamum, species: indicum. There are more than 15 varieties of sesame within Sesamum indicum. Therefore, it is possible that some of the disparities between the sesame allergens that have been found in different studies by Pastorello and were related to the search for allergens in different varieties (2,3).

Sesame seeds have a high oil content, between 50% and 60% of their weight in oil. In contrast to sunflower and peanut oil, sesame oil is always cold-pressed. It is not refined or heated, because heat would destroy the delicate taste of sesame oil. However, heated sesame oil is used in cosmetics, soap, etc.

7.2   Prevalence of Sesame Allergy

The prevalence of allergy to sesame has been appreciated prospectively (4,5), and has been estimated to be less than 1% (6). Eight years ago only data on prevalence of sensitivity in adults were given; however, there is now a databank from the Circle of Biological and Clinical Investigations for Food Allergy (CICBAA), which registers all the cases that have been studied in two

University hospitals in France - one in Toulouse (Department of Pediatrics) and one in Nancy (Department of Internal Medicine, Clinical Immunology and Allergology). The databank contains information on 1,008 observations of real food allergies that have been established by prick-tests and RAST, which are proof of sensitisation, and then by DBPCFC (double blind, placebo controlled food challenges). At the present time, this databank demonstrates a prevalence of 2.2% sesame allergy, and 3.35% in adults (7). Cases of sesame allergy in children are beginning to become apparent, as shown in the study by Levy (8). The first survey of the Allergy Vigilance Network, that was launched in 2000, revealed that 4% of life-threatening food allergies were due to sesame seeds (9).

7.3   Anaphylactogenic and Asthmogenic Indices of Allergens

The databank enables the calculation of an anaphylactogenic index of food allergens, which is the ratio of the number of anaphylactic shocks to the total number of food allergies. Anaphylactic shocks are defined by symptoms accompanied by a fall in blood pressure. Celery has a high anaphylactogenic index. Wüthrich and others are calling for celery to be added to the ten allergens that should be declared on labelling. Sesame has an anaphylactic index that comes after that for crustaceans.

The asthmogenic index is the rate of acute attacks of asthma out of a total number of food allergies, and, using this calculation, peanut is the first on the list. Sesame is another asthmogenic allergen.

There are no such calculations for laryngeal angioedema. The definition of laryngeal angioedema is a shortness of breath with modification of the voice, but it is difficult to analyse these symptoms some months after the event.

7.4   Research into Sesame Allergy

There are about 30 to 40 papers in the records about sesame allergy. A lot of papers deal with between one and three cases of sesame allergy. Two series include 10 to 12 patients (2-6-10). Our own series includes 25 patients with sesame allergy up to now[1]. It is striking that, out of 25 patients,

---

[1] Research conducted at Hopital Central, Nancy, France, (unpublished).

20 are males (seven out of 12 patients studied by Pastorello were males). This is unusual since in adults there is no such a predominance of males with food allergies.

In our series, the predominant clinical features of sesame allergy were asthma and atopic dermatitis in children. Half of the adults suffered from anaphylactic shock, with a loss of consciousness in some cases. Forty per cent of patients were not atopic. They had no previous history of atopic disease, and prick-tests to 12 current inhalants were negative.

Sensitisation is established in order to be certain that the reaction is immunologically mediated. Sensitisation up to now has been studied for IgE-dependent allergy to sesame. It is not known whether there is delayed reaction to sesame, as it is documented for milk and wheat allergies.

Results of skin prick testing to commercial extracts of food allergens are poor. Sesame extract is not reliable. Prick-in-prick tests were performed to three varieties of sesame seeds – white, brown and black varieties. Sensitivity was 80% - specificity is excellent (data not shown).

Radioallergosorbent assay (RAST) has a poor sensitivity: 40%.

7.5   Challenge Testing for Sesame Allergy

Sesame allergy was confirmed in the study conducted by Moneret-Vautrin by positive labial challenge in half of the 25[2] cases. The labial test was conducted by putting a drop of crushed sesame seed on the lip of the subject (11). DBPCFC was deemed inadvisable in young children and in adults owing to the possible severity of the reaction. Reactions of grades two, three and four were produced. A grade two reaction is the extension of urticaria on the chin; grade three is the extension of urticaria to the cheek with conjunctival hyperaemia; and grade four is systemic symptoms such as tachycardia, wheezing, etc. (11)

In 14 cases, double-blind or single-blind placebo-controlled food challenges were performed as detailed previously (12). The placebo used was mashed potato. A scale of increasing doses every 20 or 30 minutes was used, over a period of 3 to 4 hours. Patients were followed up for the next 18 hours. On the first day of challenge testing, up to 965 mg of sesame was

---

[2] Research conducted at Hopital Central, Nancy, France, (unpublished).

used, and, if responses to this level of challenge were negative, on the second day of challenge testing the dose was increased to between 7 and 10 g.

The threshold of reactivity was calculated by adding together the doses that had been ingested. One subject reacted to 30 mg, one to 100 mg, one to 265 mg, one to 965 mg, one to 3 g, one to 7 g and two to 10 g of sesame.

This is the first reported study that benefited from DBPCFC. Most papers report that, for ethical reasons, whenever the clinical reaction to sesame was severe, an oral challenge was not performed. The need for oral challenges is great. Particularly in some cases where there is no evidence of an IgE-dependent sensitisation it is the only means to be sure that a patient is allergic to sesame.

Five patients with a positive DBPCFC to sesame seeds agreed to have a second challenge a week or a month later. The cumulated reactive dose of sesame seeds ranged between 100 mg and 7 g. Symptoms included urticaria, fall of peak flow rate (PFR), abdominal pain, and angioedema. Commercial sesame oil induced a clinical reaction between 1 ml and 30 ml. Two patients suffered from anaphylactic shock after ingesting sesame oil; there was no such reaction to sesame seed. One of these two subjects had generalised erythema after using an ointment containing sesame oil for massage. The other had anaphylactic shock in a restaurant in Israel after 7 years of strict avoidance of sesame. Sesame oil in salad dressing was the probable cause.

7.6   Allergenicity of Sesame Oil

The quantity of proteins that may be included in a few ml of sesame oil is considerably smaller than in seeds. Moneret Vautrin *et al.* have performed oral challenges to peanut, sunflower, and soya oil. Occasionally, there were clinical reactions, especially with peanut oil; however, there were no cases of anaphylactic shock with oils other than sesame oil (12).

It is important to underline that, in adults, there are no examples of recovery from food allergy to sesame; it is a very long-lasting food allergy. The history of sesame allergy development in young children is not yet known.

There are many hypotheses to try to understand why such a tiny amount of protein in sesame oils is allergenic. The first consideration perhaps is that allergic patients react to very hydrophobic proteins. The importance of hydrophobic epitopes was shown for allergenicity of β-lactoglobulin (13). Allergenicity of proteins may be partly related to the intensity of hydrophobicity. Hydrophobicity could allow an easy passage through enterocytes, and also through the membrane of antigen-presenting cells. This is a mere hypothesis.

Another explanation for allergenicity of oils is that heating has modified the allergenicity of proteins, enabling interactions with lipids. The problems associated with the interactions of proteins with lipids, or with saccharides are a present-day aspect of our reflection regarding the allergenicity of proteins. However, sesame oil is said to be cold pressed, therefore heating would not play a role in modifying allergenicity.

A third hypothesis is that the lipidic nature of oil has an adjuvant effect on allergenicity.

Three out of five patients allergic to sesame oil had negative prick tests to sesame seeds and only one had a positive RAST. Two subjects had negative prick tests and negative RASTs. Some cases of severe anaphylaxis to sesame without evidence of specific IgE have been described (14). Sesame allergens are probably potent immunogens, and create a strong IgG response, as has been shown previously (15). Specific IgG could be the culprit for certain cases of IgG-mediated anaphylaxis.

The practical consequence is that it is essential to use DBPCFC in the diagnosis of sesame allergy, even in the case of negative skin tests and RAST.

When we began to study allergens in sesame, three varieties were used - white, brown and black. The total quantity of proteins is different in these three varieties: 182 mg proteins/g in white sesame, and 28 mg protein/g in black sesame. The analysis of sesame allergens was carried out using white sesame (3).

A Western blot using the serum of six patients sensitised to sesame showed mainly bands of 12 to 14 kDa; a band of 23-24 kDa confirming findings of previous studies (14,15). Many isoforms were characterised in the 32-36 kDa bands (3).

The Italian study by Pastorello clearly identified a major sesame allergen of 9 kDa in Italian sera (2). The protein was sequenced and shown to be a 2 S albumin, and shared important homologies for similarity, with the castor bean, sunflower and brazil nut 2 S albumins.

There is then a difference between allergens demonstrated by Italian sera and by French sera. An example of such a disparity concerns allergy to apple and peach. In North Europe, the major allergen is homologous to Bet v 1, the major allergen of birch pollen. Italian researchers have shown that, Italian patients allergic to apple and to peach, do not react to Bet v 1 but to a lipid transfer protein (LTP) (16). Therefore, differences in the results from the studies by Pastorello and Frémont may be due not only to different varieties of sesame used in the studies, but also genetic background and unknown environmental factors may be responsible for differences in immune responses. This finding requires further study.

7.7   Cross-reactivity

When food allergies are investigated, three levels of cross-reactivity have to be considered: cross-allergies, cross-sensitisation, shown by positive prick tests, and *in vitro* cross-reactivity.

Forty-four per cent of our patients with sesame allergy have associated food allergies. Five patients are allergic to peanut, two to paw-paw, one to kiwifruit, one to brazil nut, one to poppy seed, and one to buckwheat[3].

Associated sensitivity to tree nuts was investigated by prick tests. Out of the twelve patients tested, nine had a positive prick test to hazelnut, seven to almond, four to brazil nuts, and four to pecan nut; the prick-tests with the other nuts - chestnut, walnut, pine nut, macadamia nut, and cashew nut - were negative. The *in vitro* cross-reactivity has been shown for walnut, hazelnut, kiwi, and poppy seeds (17,18).

In conclusion, sesame allergens are potent immunogens and potent allergens. The rising prevalence of sesame allergy makes it possible that, over the next decade, sesame allergens may be considered more severe than, or at least as severe as peanut allergens.

---

[3] Research conducted at Hopital Central, Nancy, France, (unpublished).

7.8 References

1. Kimura S. Positive ratio of allergen specific IgE antibodies in serum, from a large scale study. *Rinsho Byori*, 2001, 49, 376-80 (article in Japanese).

2. Pastorello E.A., Varin E., Farioli L., Pravettoni V., Ortolani C., Trambaioli C., Fortunato D., Giuffrida M.G., Rivolta F., Robino A., Calamari A.M., Lacava L., Conti A. The major allergen of sesame seeds (Sesamum indicum) is a 2S albumin. *Journal of Chromatography B Biomedical and Applied Science*, 2001, 756, 85-93.

3. Frémont S., Zitouni N., *et al.* Allergenicity of some isoforms of white sesame proteins. *Journal of Allergy and Clinical Immunology*, 1999, 103 (1 part 2), S29.

4. Dalal I., Binson I., *et al.* Food allergy is a matter of geography after all: sesame as a major cause of severe IgE-mediated food allergic reactions among infants and young children in Israel. *Allergy*, 2002, 57(4), 362-5.

5. Sporik R., Hill D. Allergy to peanut, nuts, and sesame seed in Australian children. *British Medical Journal*, 1996, 313, 1477-8.

6. Kanny G., De Hauteclocque C., *et al.* Sesame seed and sesame seed oil contain masked allergens of growing importance. *Allergy*, 1996, 51(12), 952-7.

7. Moneret-Vautrin D. Epidémiologie de l'allergie alimentaire et prévalence relative des allergènes. *Cah. Nutr. Diet*, 2001, 36, 1-6.

8. Levy Y., Danon Y. Allergy to sesame seed in infants. *Allergy*, 2001, 56, 193-4.

9. Moneret-Vautrin D.A., Kanny G., Parisot L. Severe reactions induced by food allergies in France. *Rev. Fr. Allergol. Immunol. Clin.*, 2001, 41, 696-700.

10. Kagi M.K., Wuthrich B. Falafel burger anaphylaxis due to sesame seed allergy. *Ann. Allergy*, 1993, 71(2), 127-9.

11. Rance F., Dutau G. Labial food challenge in children with food allergy. *Pediatr. Allergy Immunol.*, 1997, 8, 41-44.

12. Moneret-Vautrin D., Rance F., *et al.* Food allergy to peanuts in France - evaluation of 142 observations. *Clin. Exp. Allergy*, 1998, 28, 1113-9.

13. Wal J. Structure and function of milk allergens. *Allergy*, 2001, 56:suppl., 67, 35-38.

14.   Eberlein-Konig B., Reuff F., *et al.* Generalised urticaria caused by sesame seeds with negative prick test results and without demonstrable specific IgE antibodies. *J. Allergy Clin. Immunol.*, 1995, 96(4), 560-1.

15.   Kolopp-Sarda M.N., Moneret-Vautrin D.A., *et al.* Specific humoral immune responses in 12 cases of food sensitization to sesame seed. *Clin. Exp. Allergy*, 1997, 27(11), 1285-91.

16.   Pastorello E.A., Vieths S., Pravettoni V., Farioli L., Trambaioli C., *et al.* Identification of hazelnut major allergens in sensitive patients with positive double-blind, placebo-controlled food challenge results. *J. Allergy Clin. Immunol.*, 2002, 109, 563-70.

17.   Alday E., Curiel G., *et al.* Occupational hypersensitivity to sesame seeds. *Allergy*, 1996, 51(1), 69-70.

18.   Vocks E., Borga A., *et al.* Common allergenic structures in hazelnut, rye grain, sesame seeds, kiwi, and poppy seeds. *Allergy*, 1993, 48(3), 168-72.

19.   Asero R., Mistrello G., *et al.* A case of sesame seed-induced anaphylaxis. *Allergy*, 1999, 54(5), 526-7.

8. FOOD ALLERGEN RISK MANAGEMENT – A MATTER OF LIFE OR DEATH?

Mrs Hazel Gowland

8.1   Introduction

As our food supply chain becomes ever more complex, people who need to avoid allergens, both as ingredients and as trace contaminants, need increasingly sophisticated and accurate information in order to make an informed food choice and prevent life-threatening symptoms. It is vital that all those involved understand every aspect of the risk assessment, management and communication involved. This work involves bridging the information gap between allergic consumers and the food industry: explaining to people who work in the food industry the realities of life as an allergic consumer; explaining to the Government what life is like for the allergic consumer; and explaining to allergic people, and their families, what is happening in the food industry and what the risks are. This is possible through consultancy, training and information.

The Anaphylaxis Campaign plays a key role in bridging this information gap. This national charity was founded by David Reading in 1994, following the death of his daughter as a result of an allergic reaction to peanut (which was unidentified as an ingredient and unexpected). The Campaign now has nearly 7,000 UK members. It supports those managing allergy risks and provides well-researched information on allergy across all aspects of life. The Campaign has regular contact with a wide range of groups, including schools, playgroups, food companies, health professionals and central and local government bodies. Its commitment to preserving and improving the quality of life for at least 1% of the population also involves initiatives to ensure that high-quality allergy patient care is available across the country, as provision in this area is currently inadequate.

The Campaign supports medical and scientific research, which will help allergic people, both to improve medical understanding, and to support the food industry to improve their management of allergic risk so that more people can make more informed food choices.

8.2 Prevalence

The most commonly mentioned allergens are peanuts, tree nuts, cows' milk, eggs, soya, fish, shellfish, wheat (the big eight) and sesame. However, the Institute of Food Science and Technology has recorded 170 food allergens; and in theory a susceptible individual could become allergic to any protein. At least one person in a hundred could suffer a potentially fatal allergic reaction to a food. It is estimated that, in school children, the prevalence of peanut allergy is 1% - that is, one child in three or four classes, which is a lot of children.

At least three-quarters of allergy deaths are due to food consumed away from the home. Current trends in eating habits and lifestyles indicate that an increasing proportion of the food we eat has been prepared away from the home.

Eighty-five per cent of the allergic consumers who join the Anaphylaxis Campaign record on their application forms that they are allergic to peanuts, and another 10% to peanuts and/or all tree nuts, which include brazil nuts, walnuts, pecans, almonds, hazelnuts, pistachios, cashews and pine nuts. People also suffer increasingly from sesame allergy: in 1995, Campaign members who had indicated that they had sesame allergy were sent a questionnaire on which they were asked to describe symptoms, the foods they thought had triggered them, and whether they had had a formal medical diagnosis. Results showed that between 8 and 9% of the membership had reported problems with sesame, including very severe reactions. One woman went into early labour following a reaction to sesame during which the baby suffered cyanosis and only just survived. Sesame is used in the UK, predominantly in children's diets to decorate burger buns, which means that its presence in fast food outlets is widespread. Some burger outlets are now decorating bread buns with grains that may already be present as flour ingredients rather than introducing sesame to an otherwise nut- and seed-free environment.

8.3   Symptoms

Symptoms of an allergic reaction are often unpredictable. They can include anaphylaxis, and allergen-triggered asthma, which can also be fatal. In some cases, people have experienced both types of symptom, and do not know how they may react when they are next exposed to allergens to which they are allergic; this unpredictability is very frightening for sufferers, and very difficult to manage. Symptoms can also differ depending on such factors as the weather, whether a sufferer has exercised, and age.

Symptoms may include the following: a swollen throat or mouth, a tickly mouth, swollen lips, and a swollen tongue, which leads to difficulty in swallowing or speaking. An allergic person recognising his/her own symptoms could be described as a "first party risk assessor". Assessing symptoms for other people or for other people's children is very different; for example, assessing the possible symptoms of an allergic reaction in a 2-year-old child is very difficult.

Symptoms also include alterations in heart rate; these are not visible to an observer, and may be disguised as panic. Skin rash, urticaria, cramps, nausea and vomiting may all be precursors to more serious symptoms and may not be easily identified as symptoms of an allergic reaction, particularly in children. An allergic reaction can also be characterised by a feeling of weakness caused by a drop in blood pressure, and a dramatic sense of impending doom; allergy sufferers report feeling a grabbing panic. Collapse and unconsciousness, followed by death, can also occur following ingestion of/or contact with a potent allergen. Death is due to a variety of causes, including suffocation, heart failure and asthma.

Treatment for an anaphylactic reaction is an injection of adrenaline – at least one, possibly more. The Campaign recommends that people who have been prescribed an Epipen or Anapen (emergency injectable epinephrine/adrenaline) actually carry more than one. In parallel with administering the injection, an ambulance should be called. Help must be brought to the person; the person suffering the reaction should not be made to do anything, or expend any energy.

In the early stages of a reaction, and while not replacing adrenaline, antihistamines and steroids can be useful. Allergy sufferers often carry these. Piriton syrup is helpful, particularly in children, although it can take some time for it to take effect. Since many allergy sufferers are also asthmatic, they may well carry rescue medication such as Salbutamol (Ventolin) (often their blue inhaler), which can be administered whist they are conscious during a reaction, and which with appropriate positioning and fresh air, can help maintain breathing.

8.4   Case Studies

The priority of the Campaign is to prevent accidental ingestion of allergens, but occasionally a real-life example of a fatal or non-fatal anaphylactic reaction that is reported in the national press does provide focus to the issue.

8.4.1   Case study 1

A young woman during a short stay in London went out to a restaurant with her boyfriend. She knew beforehand that she was nut-allergic. She ate a bread roll that contained walnuts as an ingredient, but which she did not expect to be present. The girl died after suffering an allergic reaction. Although the restaurant had identified that the bread contained walnuts on the menu, the coroner at the inquest suggested that a consumer would not necessarily look at the menu before eating a bread roll.

8.4.2   Case study 2

In a case involving a clear-cut foodservice prosecution, a girl student went to a restaurant and ordered a take-away meal. She asked whether the meal contained nuts. The restaurant said that the meal did not contain nuts. Whilst eating the meal, the customer had a serious allergic reaction, but recovered. Since she had retained the remains of the meal, it was possible to send it for analysis, during which peanut protein was discovered. This evidence enabled the local food enforcement authority (Trading Standards) to secure a prosecution under Section 14 of the Food Safety Act 1990 as the food was "not of the nature, substance or quality demanded by the purchaser", to the detriment of the purchaser. Once a dialogue has taken place between a food business and an allergic customer, and an allergy to a particular food has been declared, the onus is on the food business to ensure that the allergen status of any food sold to that customer is what the customer expects it to be. The Anaphylaxis Campaign encourages its allergic members to ask food businesses clear questions about the presence of allergens, ideally in front of witnesses, and to get the food business to make a written note of the allergen to be avoided on the order. At best this will prevent accidental ingestion of an allergen, and, in the worst situation, the written record can provide evidence of the conversation.

8.4.3   Case study 3

A student who was allergic to nuts had moved away from home to University in Durham. She had an Indian meal, which triggered serious allergic symptoms. It was reported that she had left her emergency medication in her room, and after rushing back to get it, she collapsed and died. Although legally adult, such young people moving out into the wider world are particularly vulnerable. They may have more control over their actions, but may not have discussed their allergy with new colleagues or friends. The Anaphylaxis Campaign works with the National Union of Students to raise awareness of potentially fatal allergies across campuses at the beginning of each new academic year.

8.4.4   Case study 4

A young woman died after eating a muffin from an in-store bakery. It was suggested that the product was sold in a bag that carried a warning about possible nut trace contamination. Products from in-store bakeries can be very dangerous for nut and seed allergy sufferers because of the serious risk of cross-contamination from other unwrapped products. The Anaphylaxis Campaign advises those at risk from these allergies to avoid bakery products sold loose, or from local and in-store bakeries. It was alleged in this case that a reduced price sticker covered the "may contain" warning on the product.

8.4.5　Case study 5

A young athlete who was nut-allergic had been training, and asked someone with him to get a chicken sandwich. This person bought a coronation chicken sandwich; the Coronation was in 1953 and this incident took place in 1998; Many children and young people have little formal food education; food preparation has a very minor place in the National Curriculum, and people who were not around in 1953 may not necessarily know that coronation chicken can have nuts in the recipe. This is one of many cultural and linguistic dilemmas in allergen risk assessment and communication The athlete consumed the sandwich, had an anaphylactic reaction, probably to the nuts which were ingredients in the coronation chicken filling, and, after some days on a life-support system, he died.

8.4.6　Case study 6

A student at Cambridge University attended a college dinner at the beginning of her first term. As a young fresher, she may not have wished to draw attention to herself by asking waiting staff exactly what was in the food. The menu listed "strawberry shortbread on a red fruit sauce." The student collapsed after eating the meal and died some days later. It is believed that the dessert was actually "strawberry and almond shortbread on a red fruit sauce". If this description had been on the menu, the student may have still been alive: she knew she was allergic to nuts, and might have avoided the dessert.

8.4.7　Case study 7

Another case of anaphylactic reaction involved a girl who was working as a waitress at the air show at Farnborough. In her break she ate a fruit tart, which had fruit on the top and pastry on the bottom, but hidden in between was a layer of frangipane, which contains almonds. She had an allergic reaction, and when more serious symptoms developed paramedics were called in. In this case the allergy sufferer survived because medical help was readily available at this public event.

8.4.8   Case study 8

In a case of a domestic mistake, a boy who was allergic to cows' milk had a special cup for his soya milk. A visiting relative did not know this and put cows' milk in his cup. The boy died of asthma, triggered by the food allergen.

8.4.9   Case study 9

A 14-year-old boy who was cows' milk allergic was at a summer sleep-out in a garden with his friends. The boy died as a result of a reaction to whey powder, which was an ingredient in the flavouring of crisps which he ate.

In circumstances such as these, young people may be excited, and possibly not as vigilant as usual about what they are eating; they are also perhaps eating things that they would not normally have. This reaction occurred in high summer, when allergic people may be subject to seasonal allergies such as hay fever, which may exacerbate further food-triggered symptoms.

8.5   Key Risk Factors for Allergic Reaction

The key risk factors for a severe reaction are: the food is usually prepared and eaten away from the home; the person is between 13 and 25; and it is usually his/her first experience of a serious reaction. Teenagers in particular may not maintain preventive therapy for asthma (usually a brown steroid inhaler); it is possible that their asthma may have improved in adolescence, and they may try to manage without the medication that their parents insisted they took as children. One outcome is the removal or reduction of a certain threshold of protection from severe allergic symptoms.

In many cases, allergic reactions are a result of misunderstanding or confusion about what might be in a food product; for example in the case of the athlete who died after eating a coronation chicken sandwich - the allergen was present quite deliberately, but he did not know this. There are many instances where an allergen is present in a food, but is unrecognised, and unexpected by an allergy sufferer.

In many cases of fatal anaphylactic reactions, there had been no formal specialist medical diagnosis, dietary guidance on allergen avoidance or emergency rescue treatment available. The current provision for medical allergy testing in the UK is currently inadequate at a time when the prevalence of allergy appears to be increasing.

8.6   The Role of the Food Industry

A personal experience at the age of 18 involved a near-fatal reaction to a chocolate caramel of about three centimetres square from a selection box. The chocolate coating was milk chocolate and a very serious reaction developed within 35 minutes; however, adrenaline was administered, and slowly the extremely dangerous symptoms started to reverse. The chocolate coating on the caramel had 1% rework in it, and this incident demonstrates the low level of contamination that can cause a serious allergic reaction.

Many manufacturers and retailers now recognise that the food industry carries some responsibility to prevent incidents such as this, and are increasingly involved in allergen risk management, assessment and communication activities.

The Anaphylaxis Campaign is involved in many initiatives with UK manufacturers, retailers and the foodservice sector and has contributed to guidelines on Good Manufacturing Practice with respect to allergens, produced by the Institute of Food Science and Technology, and Voluntary Labelling Guidelines, available from the Institute of Grocery Distribution.

The Campaign is attempting to incorporate training on food allergy and intolerance into measures being taken for other initiatives within the food industry – for example, hygiene training and risk assessment. Many UK food retailers, manufacturers and some caterers are now engaged in HACCP (Hazard Analysis and Critical Control Points) risk management. In many cases, there is significantly better traceability of food allergen risks in the UK than on the Continent. UK food retailers are in a position of influence and play a useful part in encouraging their suppliers to identify food allergen risks, to segregate food allergens where possible, and to communicate allergen risks to the consumer.

Detection methods for allergens are being developed and employed more widely. Whilst there are commercially available assays for some of the most common allergens, new tests for additional allergens (e.g. a wider range of nuts) and their improved reliability are both vital to support and demonstrate the effectiveness of HACCP allergen management procedures. Current initiatives include new developments in product traceability using specialised software. This improves the availability of product data (including the allergen status of a food) for the full length of a very complicated food supply chain. Such developments are important in identifying the risk that a particular product may pose to an allergy sufferer: a pre-prepared meal may contain ingredients from several different countries, and information on contact with allergens must be available from every supply source.

There are initiatives in place to improve access to foodservice product information, particularly in the kinds of place where children eat, and to provide information to the customer. In collaboration with efforts made to inform consumers about possible allergens in foods in catering situations, consumers must themselves be educated about their food allergy.

The increase in multicultural dining, development of new recipes, and product innovations are all challenges to the task of informing and safeguarding food allergy sufferers, together with the practical difficulties of managing allergens in the catering situation. In many cases, space is limited. Product information from suppliers, on packaging labels, held by kitchen and service staff, and on menus has to be accessible and reliable. It is important to try to regulate this environment so that allergic people (and their colleagues, friends and family members) are able to eat in as many places as they possibly can and to make informed food choices.

The perception of managing life while suffering from food allergy, and the reality of the situation are possibly not quite the same. Parents of younger children are anxious; in many cases they are terrified – they may not have had to consider such an immediate and potentially serious issue before, and many cannot see any further than the very short term. There is a huge burden of risk assessment, communication, panic, and guilt regarding the vigilance required, carrying emergency medication and worrying about what friends and other family members might give the allergic child to eat. This level of stress can cause serious social exclusion for otherwise very normal children. For example, in exceptional cases, children have been sent out of school at lunchtime because there is inadequate supervision from support staff. In such circumstances, when children are eating, when there may be insufficient adults present to supervise, staff in some schools have decided that the risk is too great and have sent the allergic child home. Allergic children have been prevented from joining play schemes, and other extra-curricular activities. The only way to manage such a situation is by improving communication, and developing good relationships with those who come into regular contact with an allergic child.

There is a current trend towards pre-prepared food that requires minimum preparation at home. Consumption patterns such as this force the consumer to rely on other people's risk management in the preparation of their food. Consumers are also encouraged to enjoy a huge variety of food choices - to be able to eat food from many different cultures that were not part of the mainstream diet in the UK 10 to 20 years ago. Consumers also have less time, and are unprepared or are unwilling to cook; in fact, in some cases it is cheaper not to cook. Fewer people understand what is in food, partly because they are not learning about it in school, and partly because they are buying items that are pre-packed or prepared freshly for them, and are not cooking at home.

8.7   Allergen Labelling

Life-saving information to indicate whether a product contains an allergen is very often difficult to find on product labelling; it can be printed in a very small typeface - or hidden perhaps under the flap on a packet. Labelling is also occasionally printed on a surface such as transparent wrapping of a sandwich, which makes the text very difficult to read. In contrast, the labelling on a packet of cigarettes indicating that smoking is harmful to health is very prominent. The immediate health effects of smoking are negligible, whereas the symptoms triggered by ingesting an allergen may become critical within minutes.

If an allergen is included in the formulation of a food because it is essential to the character of the product, and the allergen is added deliberately, the product can be highlighted as "containing" the particular allergen. Ideally, on product labelling, there is a list of the ingredients used, and a statement regarding which allergen(s) the product contains – both as ingredients and as possible trace contaminants. This rule is followed by most of the retailers in the UK for the big "8" allergens (peanuts, tree nuts, fish, crustacea, soya, wheat, egg and milk); sesame is also included in this voluntary practice.

In order to help allergic consumers, labelling using "may contain traces......" or similar wording should be used only as a short-term measure. In the medium and longer term, the use of this type of labelling referring to known allergens is not acceptable because serious food allergies now affect so many people. For example, peanut allergy affects one in a hundred children for whom symptoms may be unpredictable. Use of "may contain" labelling should follow the implementation of a proper risk assessment, and there should be a measured and a recorded risk of allergen trace contamination. The responsible manufacturer should have done a full assessment of all his suppliers. If an allergen contamination risk has been recognised and cannot be eliminated in the short term, - i.e. the ingredient cannot be sourced from elsewhere – a warning is used on labelling. It is important that the allergen risk is communicated effectively to consumers, because it is life-saving information.

The use of "free from....." labelling can be very helpful in the short term, particularly when confectionery, biscuits and cereals are free from nuts and peanuts. If it is certain that a product is "free from" nuts, this labelling is helpful on foods that otherwise might well be considered to put the allergic consumer at risk. It is not helpful on products that would appear to have no obvious connection with, or contain nuts (for example), such as bags of flour or pints of milk. The Anaphylaxis Campaign supports such 'free from' labelling only as a very temporary solution. Allergic consumers are advised to rely on ingredients labels as their primary information source.

Young adult consumers in particular find this frustrating enough and will spare no further time or effort to find a 'free from' or 'may contain' warning label. In the medium and longer term, full segregation of known allergens will mean that an allergen not mentioned as an ingredient will not be present – even as a trace contaminant.

8.8   Aims of the Anaphylaxis Campaign

The Anaphylaxis Campaign's aims include:

- A reduction in the use of "may contain labels", once the risk of allergen contamination has been assessed and eliminated.
- Clear and legible ingredients information as a priority.
- Allergen labelling that is easy to find, read and understand
- Allergy training for caterers integrated into the rest of their professional and hygiene training.
- Food hygiene enforcement, integrating allergy risk into HACCP so that it is included with all the other factors considered when establishing a HACCP plan
- Improved availability of high-quality patient care

The Campaign runs workshops for high-risk allergy sufferers - older primary children, teenagers and young adults. The workshops encourage people in this high-risk group to check everything that they eat, assertively and confidently, without parental help, and how to handle situations that arise in real life, such as remembering to carry emergency medication when they go clubbing, and how to handle eating in restaurants and take-aways. These and other initiatives help to ensure that all those at risk from serious food allergies can live normal lives with confidence and competence.

9. DO FOOD ADDITIVES CAUSE HYPERACTIVITY?

Dr Taraneh Dean

9.1   Introduction

This paper covers the results of a study conducted on the Isle of Wight between 1997 and 2000, which looked at the association between food additives and behavioural problems - hyperactivity in particular (1).

Hyperactivity is actually much more prevalent than one might think and is a commonly diagnosed behavioural disorder, particularly in a primary care setting. The symptoms are varied, and include difficulty in concentrating, an excessive level of activity, and impulsive behaviour. These symptoms are frequently found in young children who may not suffer from the disorder.

Comorbidity has been researched, and there is good evidence to suggest that those children who have a diagnosis of hyperactivity and attention deficit hyperactive disorder (ADHD) have difficulties in language development, and in adapting to school life, and academic achievement (2,3). In addition, there are many confounding factors, which could influence any investigation of hyperactivity. These could include diet, social class, gender (particularly male) and possibly allergy and atopy.

9.2   Prevalence of Hyperactivity

The prevalence of hyperactivity has been discussed and cited widely in the literature. Prevalence depends very much on the diagnostic criteria used: some reports suggest that prevalence of hyperactivity is about 2%, and others suggest that it is nearer 18% (4,5).

There is particular interest in the influence that diet has on hyperactivity; however, previous studies have had methodological problems. There is very poor case definition, and diagnostic criteria vary from country to country, and clinic to clinic (6).

Most studies in this area are very much dependent on the subjective rating of behavioural change. The study by Dean *et al.* (1) utilised a combination of both subjective and objective rating.

9.3   The Study

9.3.1   *Objectives*

The study by Dean *et al.* (1) set out to establish the prevalence of hyperactivity and behavioural problems in a population of 3-year-old children. In addition, the prevalence of atopy as indicated by sensitisation as well as symptoms of allergy were also studied in this group, and any association between those children diagnosed with hyperactivity and those with allergy and atopy were investigated.

The study investigated the effect of artificial food colouring and preservatives on children's behaviour, and any trends in the ways in which children were affected by these substances and whether the atopic, or allergic children were more prone to these influences.

9.3.2   *Study methodology*

The study was conducted in three phases. Phase one was to establish the prevalence of hyperactivity. Validated questionnaires were used, which were administered mostly over the telephone by the research nurse on the study.

Phase two of the study looked at prevalence of atopy, using skin prick testing as a marker of sensitisation. Symptoms of atopy were studied using the questionnaires from the International Study of Asthma and Allergy in Childhood (ISAAC), which is a huge collaboration involving many countries.

Phase three of this study was a double-blind placebo-controlled (DBPC) cross-over trial with artificial food colourings and sodium benzoate.

All children who lived on the Isle of Wight and were 3 years old over a 2-year period were screened with the validated questionnaires for hyperactivity, which formed phase one of the study. At this point, the children were identified as either hyperactive or not, and, if they were not hyperactive, they were put in the control pool.

Hyperactivity was assessed using two validated questionnaires, which have been used in previous studies to assess hyperactivity. One questionnaire was the "Emotional Activity and Sociability Temperament" questionnaire (EAS); the other was the "Wise-Werry and Peters" (WWP) activity scale (7).

Behaviour was assessed using a validated behavioural checklist (8). Parents were also asked whether they perceived their child to be hyperactive and whether they thought that food or drink played a role in the hyperactivity. The child was defined as hyperactive if he/she scored 20 or more on the WWP questionnaire and an average of 4 or more on the EAS questionnaire. Equally, if children had a score of 10 or more on the behavioural checklist, they were scored as having a behavioural problem.

The children were screened for atopy by skin prick test in phase two. In addition, the clinical research fellow examined the children for signs of allergy and completed the ISAAC questionnaire for allergic symptoms. The children were then classified as either atopic or not atopic.

Phase three involved four groups of children: those who were hyperactive as well as atopic; those who were hyperactive but not atopic; those who were atopic but not hyperactive; and those who were neither hyperactive nor atopic.

It was hoped that, for every hyperactive and atopic child, there would be at least one child in the other cells matching them for sex and for social class.

The children who took part in the study were born between 1994 and 1996 and were resident and registered with a GP on the Isle of Wight (n=2878). Screening with behavioural questionnaires was completed on 1,873 children. Subsequently, 1,246 children went on to take part in phase two and completed the skin prick tests. A total of 277 children completed phase three.

9.3.3 Results

9.3.3.1 Phase 1

The prevalence of hyperactivity amongst the 3-year-olds studied was 16.4%. It is important to bear in mind that the population studied was not specially selected; every 3-year-old who lived on the Isle of Wight was approached, and the majority of the available target population participated in this phase of the study.

The prevalence of behavioural problems was 23.4%. The prevalence of 16.4% for hyperactivity was generally in line with other studies that have looked at hyperactivity in this age group. However, the level of behavioural problems did seem slightly higher than what had previously been reported. The usual reporting of behavioural problems for this age group is between 16 and 17%; therefore, the population studied on the Isle of Wight had a higher prevalence of behavioural problems than normally found. It is not known why this was the case.

On breaking down these prevalence figures in more detail and analysing them according to gender (because there is good evidence (9) to support the theory that there may be gender differences in hyperactivity), it was found that boys were generally significantly more hyperactive than girls. Equally, for behavioural problems, there were significantly greater behavioural problems reported for boys than for girls.

One-third of the parents of children who took part in the study believed that food and drink affected their child's behaviour. There was no gender difference in this case; however, one-third was quite a significant proportion of the whole study population. This result correlates with that of the High Wycombe Study (10), in which the perception of food and drink related to the problem of hyperactivity was around 25%.

Asking parents whether their child was hyperactive was a good test of finding out whether the child was not hyperactive. Asking a parent about their child's hyperactivity had a very high specificity, of 91%; however, parents were not good in determining whether their child was hyperactive.

9.3.3.2 Phase 2

The ISAAC questionnaire has been validated for 6- to 7-year-olds, but not for the age group used in Dean's study; however, it has been widely used for all children, as well as younger children.

Symptom reports were collected on rhinitis, hay fever, eczema and asthma; however, only results on eczema and asthma are reported here. Respiratory symptoms are notoriously difficult to diagnose in small children and, in order to assess the presence of respiratory symptoms in the children in the study, their parents were asked whether their child had ever wheezed, whether they had wheezed in the last 12 months, and whether they had had a diagnosis of asthma by a clinician. Results showed that there was a difference between boys and girls, with boys were given a diagnosis of asthma much more frequently than girls. There was no gender difference for ever wheezing, or wheezing in the last 12 months, but quite a large proportion of the parents of children in the study reported lifetime wheezing.

There was no gender difference with respect to the prevalence of eczema. When parents were asked whether their child had ever had eczema, over 40% reported that their child had had eczema, but when more specific questions were asked regarding the occurrence of eczema, this incidence dropped quite dramatically. This finding could mean either that parents are much more at ease in labelling their children's rash as being eczema, or clinicians may actually label a rash far too quickly as eczema.

Atopy was defined by skin tests and allergic symptoms. It was not surprising that allergic symptoms were present in significantly more atopic children than non-atopic children. Of the 1,200 children who had a skin prick test, 236 had a positive result. This gave a point prevalence of atopy of about 19%, which is similar to other studies (11) that have looked at this age group. Twenty-three per cent of the atopic children had no allergic symptoms, and 25% of those who had allergic symptoms were not atopic.

On investigating the skin prick test data in more detail, of the 236 children who had a positive result, house dust mite was the most common allergen. Peanut was the food that caused the greatest number of positive tests, in contrast to what were previously thought to be the main allergens in this age group - milk and egg.

9.3.3.3 Phase 3

Children who went through phase three of the study visited the clinic five times during a period of just over a month. Children attended at week one to have a psychological test, to see the dietitian on the study and to be given advice to go on an additive- and preservative-free diet for the period of phase three, then return one week later to have the same assessment done again; this was the baseline assessment. At this time, the children were randomly allocated into groups and received either an active drink cocktail, or the placebo. The dietitian prepared the drinks, and the rest of the research team as well as the parents/child were blind as to whether the drinks contained the additives/preservatives or not.

Development of a palatable fruit cocktail, which was liked by children yet indistinguishable between the placebo and control was not an easy task. The fruit cocktail underwent extensive pilot testing for palatability. Children were given seven bottles of drink each containing 300 ml. One bottle per day was consumed during the treatment periods. The active drink contained a total of 20 mg of artificial food colourings, and 45 mg of sodium benzoate. This was equivalent to the total colouring and preservatives used in five tubes of smarties. The level of benzoate was equivalent to that of six glasses of diluted squash. Children returned to the clinic a week later after consuming either the active or placebo drinks for the same psychological assessment as conducted at the beginning of phase 3. This was followed by a wash-out period, where no fruit drink was given, and on return to the clinic another assessment was conducted and either the active cocktail or the placebo was given, depending on the previous treatment group. The final visit after this second dosing period was visit five.

Thirty-six hyperactive atopic children completed this phase and there was a drop-out rate of 30%. Children that were atopic and hyperactive were the most difficult to find: when the study started it was hypothesised that there would be many children who were hyperactive as well as atopic. However, clearly this was not the case. A large number of children were atopic, and a fair number were hyperactive, but there were not many children who were atopic as well as hyperactive, which cast doubts on associations between these two conditions. The evidence that exists on the two conditions being linked is always based on a very select cohort of children. This population-based study could not substantiate this claim.

More boys than girls completed phase three of the study. It was no surprise that the number of children that had behavioural problems was higher in the two study groups identified as "hyperactive". Factors such as mean maternal age and leaving full-time education were used as a means of assessing social class, and there were no great differences between the four groups in this respect.

During the time that children visited the clinic during phase three, they had a period of free play, during which they were observed for attention span and, amongst other things, the frequency with which they changed the toys that were available to them was observed. A delay aversion task was conducted, which involved hiding stickers, where the children had to wait for a certain length of time before they could find the stickers. Impulsivity was tested using a test known as "Bear and Dragon Task", which is similar to "Simon Says", and the child must do what the bear tells them to do and not what the dragon tells them to do.

The children's attention and level of activity were tested by getting them to walk slowly on a line and to draw slowly to join two pictures on a piece of paper. During the entire time of doing the clinical test at each visit they wore what is known as an Actometer, which measures the amount of movement of the child - one of the symptoms of hyperactivity.

Parents were also asked to keep diaries on activity, level of movement, and whether they felt their child behaved particularly badly in a particular time period. This parental score was used in addition to these objective clinical tests.

The tests conducted at the clinic showed no evidence for any changes over the time period of testing. This was true for both active-then-placebo as well as placebo-then-active groups. One explanation for this is that there is an element of behavioural conditioning: the children came into the clinic and knew they were going to have one-to-one attention and it was a game that they enjoyed playing and as a result behaved much better.

The parental rating was much more sensitive to changes in behaviour than psychological administered tests.

The parent ratings indicated a reduction in hyperactivity between baseline and time 1 (the additive-free period). In the active-then-placebo and the placebo-then-active groups there were increases in hyperactivity for both the placebo and active challenge periods. However, the gradient of the data (the degree of change) indicated a greater increase in hyperactivity during the active periods. There was a decrease in hyperactivity reported by parents in the wash-out period between challenges.

The key finding of the study is that the changes in hyperactivity were independent of the child's atopic status, or whether the child was initially at the extreme level of hyperactivity or not hyperactive at all. Also, there was no gender difference in this result.

The effect of removing colorants from children's diets was to reduce hyperactivity in children diagnosed as hyperactive, from 15 to 6%. This is less effective than Ritalin, which is used for treatment of ADHD and has a similar effect to Chlorodin, which is sometimes used for the ADHD and Turret's Syndrome.

The study found significant changes in the hyperactivity behaviour of children produced by removing colorants and food additives. Future work would be to replicate this study in other population samples, and different groups of children, possibly older children.

9.4   Acknowledgements

The author acknowledges the research team who conducted this project. These include the clinical research fellow, Dr Belinda Bateman, the research nurse, Jane Grundy, and the research dietitian, Carol Gant.

The study was conducted in collaboration with two academic collaborators, Professor Jim Stevenson, Professor of Psychology at Southampton University, and Professor Warner from the Child Health Department at Southampton University.

In addition, the author would like to acknowledge the Food Standards Agency for funding this study.

9.5   References

1.   Do food additives cause hyperactivity and behavioural problems in a geographically defined population of 3 year olds? Funded by Food Standards Agency, 1997, Project code: 07004.

2.   Merrell C., Tymms P. Inattention, Hyperactivity and Impulsiveness: Their impact on academic achievement and progress. *Br J Edu Psychol*, 2001, 71 (Pt 1), 43-56.

3.   Mannuzza S., Klein R.G., Bessler A., Malloy P., LaPadula M. Adult outcome of hyperactive boys. Educational achievement, occupational rank, and psychiatric status. *Arch Gen Psychiatry*, 1993, 50, 565-76.

4.   Taylor E., Sandberg S., Thorley G., Giles S. *The Epidemiology of childhood hyperactivity*. Maudsley Monograph. London, Oxford University Press. 1991.

5.   McArdle P., O'Brien G., Kolvin I. Hyperactivity: prevalence and relationship with conduct disorder. *J Child Psychol Psychiatry*, 1995, 36, 279-303.

6.   Lahey B.B., Pelham W.E., Stein M.A., Loney J., Trapani C., Nugent K. *et al.* Validity of DSM-IV attention-deficit/hyperactivity disorder for younger children. *J Am Acad Child Adolesc Psychiatry*, 1998, 37, 695-702.

7.   Routh D. Hyperactivity, in *Psychological management of paediatric problems*. Ed. Magrab P. Baltimore, University Park Press. 1978, 3-8.

8.   Richman N., Stevenson J., Graham P.J. *Pre-school to school: a behavioural study*. London, Academic Press. 1982.

9.   Gaub M., Carlson C.L. Gender differences in ADHD: a meta-analysis and critical review. *J Am Acad of Child Adolesc Psychiatry*, 1997, 36, 1036-45.

10.   Young E. *et al.* A population study of food intolerance. *Lancet*, 1994, 343, 1127-30.

11.   Tariq S.M., Matthews S.M., Hakim E.A., Stevens M., Arshad S.H., Hide D.W. The prevalence of and risk factors for atopy in early childhood: A whole population birth cohort study. *J Allergy Clin Immunol*, 1998, 101, 5, 587-93.

10. LACTOSE INTOLERANCE: A SUMMARY

Dr Phillipe Marteau

10.1   Introduction

Lactose is a sugar that is naturally present in milk. It is a disaccharide consisting of glucose and galactose, and, because milk is used as a constituent of many products, it is also present in milk derivatives. Lactose cannot be absorbed directly to the intestine; it must be broken down by β-galactosidase (lactase) into its constituent monosaccharides.

Table 10.I shows the lactose content of some food products; milk contains high levels of lactose, but other products, such as cream and cheeses, contain slightly less.

TABLE 10.I
Lactose content of foods

	g/100 g
Milk	4-5
Yoghurt	5.2
Cream	3.1
Gruyère cheese	2.9
Camembert cheese	Trace
Butter	0.4

Yoghurt contains a high level of lactose because, during yoghurt manufacture, lactose is digested to produce lactic acid; however, milk powder is added at the end of the fermentation process, raising the overall lactose content.

10.2   Lactose Digestion

Lactase is the enzyme required to digest lactose, and is present in every newborn mammal. Lactase is present in the brush border of enterocytes, and there is a high level of lactase activity in newborns. However, what is usually observed in most mammals is that there is a decline in the lactase activity after a few years; therefore, the majority of adults have low lactase activity. This is genetically determined, and, even if large quantities of lactose are ingested regularly, this genetically determined decline cannot be avoided. This phenomenon is called adult-type hypolactasia.

In humans, there are some exceptions to this rule: a proportion of humans maintain a high lactase activity - for example, in France, 50% of the adult population retain a high lactase activity. However, this is not the norm; in the world in general, 95% of people have low lactase activity as adults.

Other reasons for low lactase activity are not only the genetic predisposition to low lactase activity, but intestinal disease characterised by lesions of the enterocytes - for example, coeliac disease or enteritis. Intestinal lactase also requires time to act upon lactose in the gut and, if there is very rapid intestinal transit, lactase may not have enough time to digest lactose present in food.

10.3   Lactose Malabsorption

In a situation where lactose is not digested (and this is not just the case with malabsorbed lactose, but also with other non-digestible oligosaccharides, e.g. lactulose, etc.), there is an osmotic effect in the small and large bowel before digestion by bacterial lactase in the large bowel. When fermentation occurs in the large bowel, there is an increase in bacterial mass, and production of short-chain fatty acids, which results in a decrease of colonic pH. Gas production is also a feature of lactose fermentation, including production of H_2 and CO_2, and in some subjects there is also production of methane.

The results of the fermentation of some non-digestible oligosaccharides are an increase in bifidobacteria and butyric acid; however, this is not usually the case with lactose. The osmotic effects and fermentation of non-digestible oligosaccharides may produce beneficial effects; however, there may also be side effects, which are manifested as intolerance.

The use of lactose in food is frequently avoided, and new oligosaccharides such as lactulose are used; however, these oligosaccharides have the same effects as lactose: they are fermented in the colon and increase bacterial mass, starch and fatty acid production, so this strategy is not effective in reducing these side effects.

10.4   Lactose Intolerance

Symptoms of lactose intolerance are related to the dose ingested. Diarrhoea is a symptom of intolerance with large doses of lactose, and there is a threshold effect for this symptom. Prior to development of diarrhoea, an intolerance sufferer will experience increased gas production and noise in the bowel (borborigmi), followed by pain. There is no threshold effect for gas production from lactose ingestion because all humans produce an amount of gas in their gut.

The symptoms of lactose intolerance are non-specific, and are characteristic of bowel disease, the most common one being irritable bowel syndrome (IBS), which is found in 15% of the adult population, and is more common in women than in men.

Diagnosis of lactose intolerance is not straightforward. Lactose maldigestion can be determined by H_2 breath tests: a fasting subject is given a dose of lactose, and H_2 excretion is measured in breath; an increase in H_2 production indicates that lactose has not been absorbed and has reached the colonic flora. Observing the effects of maldigestion is not sufficient to diagnose lactose intolerance: the symptoms of lactose intolerance should be demonstrated to stop when lactose is removed from the diet.

A paper reviewing research into lactose intolerance was published in 1975, and clearly showed that less than 20% of subjects with lactose malabsorption demonstrated intolerance when they ingested 12 g lactose in the fasting state, which is the equivalent to two glasses of milk (1). The fasting state is the worst time for toleration of lactose. Therefore, lactose intolerance is not a frequent event, even in patients with malabsorption.

Self-reported intolerance is also known to greatly overestimate real intolerance.

In a study by Vesa *et al.* (2), 39 subjects with lactose malabsorption were given lactose at doses ranging from 0-7 g. There were no differences in symptoms between subjects who consumed 7 g of lactose and subjects who consumed no lactose; all subjects demonstrated symptoms of malabsorption. This demonstrated a placebo effect whereby people reported symptoms without ingesting the substance thought to cause the symptoms; there is a large psychological element to manifestation of these symptoms.

The true situation with lactose intolerance is the relationship between the amount of lactose ingested and the symptoms shown: the more lactose ingested, the more lactose will be malabsorbed and the greater the symptoms will be. As was demonstrated with low levels of lactose ingested, below 7 g, there is no difference in non-specific intolerance.

10.5   Lactase Activity

There has been a lot of research conducted on whether lactase activity can be modified with food, drugs, or genetics. Is it possible to stimulate the genus lactase, or add new lactase to the gastrointestinal tract, or can the endogenous lactase be allowed more time to work by slowing gut transit time? All three options are possible.

There has been one study in humans that demonstrated that it might be possible to increase lactase activity (3). Seven humans with low lactose activity ingested a probiotic, the yeast *Saccharomyces boulardii*, for 2 weeks. An increase in the endogenous lactase activity of intestinal biopsies was observed. The effect on lactase activity was thought to be due to polyamines that were present in the yeast, which may have stimulated the perforation of the mucosa and particularly lactase activity in the duodenum. However, this study was done in isolation and was not repeated, and the effect of the increase in intestinal lactase activity on lactose digestion was not examined.

It has also been shown that ingestion of yoghurt can increase the mucosal lactase activity in rats. However, when this theory was tested in humans, there was no modification of the duodenal lactase activity in people ingesting yoghurt regularly for 2 weeks (4).

Another way to increase lactase activity is to eat lactose. The vehicle containing lactose is also important as malabsorption is greater with milk than with yoghurt. Products containing lactase-containing bacteria are better tolerated than milk.

In a study by Marteau et al. (5), breath H_2 was examined after ingestion of lactose in the form of milk, heated yoghurt, or unheated yoghurt by subjects with adult-type hypolactasia. H_2 excretion rose after ingestion of milk containing 20 g of lactose, indicating lactose maldigestion. On the second day of the study, 20 g of lactose were consumed in yoghurt, and the level of malabsorption of lactose was far less than that after consumption of milk. Consumption of heated yoghurt resulted in greater malabsorption of lactose than occurred after consumption of unheated yoghurt because lactase was destroyed by heating.

On another day, the subjects swallowed a long tube and chyme was sampled at the end of the small bowel in the ileum when subjects ingested 400 g of yoghurt, which contained 18 g of lactose. The same sampling of chyme from the ileum was conducted after consumption of 18 g of lactose in heated yoghurt. There was almost no lactase activity at the end of the small bowel when the subjects ingested heated yoghurt; however, when yoghurt containing live bacteria was ingested, a reduced flow rate of lactase was observed. The lactase was from the yoghurt ingested; approximately one-fifth of the lactase contained in a cup of yoghurt could survive to the end of the small bowel. This may explain why the flow rate of lactose was lower with yoghurt than with heated yoghurt. Eighteen grams of lactose were ingested in total, and, at the end of the small bowel in those subjects with lactose malabsorption, 1.8 g of lactose was collected after consumption of yoghurt, and 2.9 g of lactose were collected after consumption of heated yoghurt.

This phenomenon of lactose toleration being dependent on the dose vehicle has also been observed in patients with short bowel syndrome.

Arrigoni *et al.* (6) studied 14 adults who had short bowel syndrome with the mean length of the small bowel being 60 cm. Subjects were given a 20-g lactose load in the form of milk and yoghurt and it was observed that the digestibility of lactose was 50% from milk and 76% from yoghurt.

10.6 Bacterial Lactase Activity

Many people have studied bacterial lactase activity, and it is not always efficient. For example, if yoghurt bacteria, or other bacteria, such as *Lactobacillus acidophilus, Bifidobacterium*, etc. are added to milk, just prior to ingestion, lactose absorption is not improved (7). Addition of extra quantities of lactase to yoghurt does not increase lactose absorption (8); and altering the lactase content of yoghurt does not improve lactose absorption as it is already at optimum in plain yoghurt (9).

In a study by Martini (10) malabsorption was quantified in subjects with lactose malabsorption using milk, yoghurt and milk containing different bacteria that were added just before ingestion of the milk. The yoghurt bacteria *Streptococcus thermophilus* and *Lactococcus bulgaricus* were not as efficient in digesting lactose when they were added to milk as when they were present in yoghurt. Bifidobacteria and *L. acidophilus* were less efficient in digesting lactose. The best bacteria for lactose digestion were found to be *Lactobacillus bulgaricus* and *Streptococcus thermophilus*, because these bacteria are very sensitive to acid and bile; they die very easily in the upper part of the small bowel, and when they die they release lactase activity. Those bacteria that are much more resistant to bile retain lactase activity inside the cell, and it is more difficult to digest lactose when the lactase is still inside the cell.

The bacteria that die easily in the small bowel, releasing lactase, can be used as vectors for other enzymes or active principles in the gastrointestinal tract. For example, *Lactococcus lactis*, which dies in the upper part of the small bowel, has been modified to produce interleukin 10 (IL-10), which is a very active ingredient in treating colitis (11). This genetically modified *Lactococcus lactis* was efficient in the treatment of mice with colitis. A recent study used *Lactococcus lactis* that was genetically modified to produce lipase. The genetically modified *Lactococcus lactis* was introduced into pigs with pancreatic insufficiency, and was shown to help the digestion of lipids in those pigs (12).

10.7 Gastrointestinal Transit

If gastrointestinal transit is fast, there is not enough time for lactose to be digested, and if gastrointestinal transit is slowed down, there is more time for digestion, and malabsorption and symptoms of malabsorption are reduced. This has been proven with loperamide, which slows down intestinal transit, and the symptoms of intolerance are reduced.

Several studies have shown that the digestion of lactose from whole milk is better than that from skimmed milk (13,14). More lactose malabsorption is observed with skimmed milk than with whole milk because the fat content of whole milk causes the gastric emptying time to be slower, which allows better digestion. The same effect is observed when cocoa or fibre is added to milk (15,16). Another study examined the effect of eating food together with a source of lactose and found that, when a sandwich was eaten at the same time as milk was consumed, the lactose present in the milk was digested more efficiently (17).

Mahé demonstrated that the half-time for gastric emptying of milk was 24 minutes; however, the half-time for gastric emptying of yoghurt was 70 minutes (18). The yoghurt was digested as semi-solid food, allowing more time for digestion of lactose. This is the second reason why lactose from yoghurt is very well absorbed: there is lactase present, and slow gastric delivery.

Colonic flora is also very important for lactose tolerance. There is colonic adaptation to some sugars that are not digested in the small bowel, such as non-digestible oligosaccharides, and this has been very well studied with lactulose, which is very similar to lactose. When lactulose is ingested regularly, changes in the colonic flora have been observed; there is a decrease in colonic pH and, when colonic pH is lowered, H_2 excretion is reduced.

10.8   Adaptation to Chronic Lactate Consumption

In a study by Briet (19), a 20-g lactose load was given to subjects with lactose malabsorption, and then they were treated for 4 weeks with lactose, or sucrose as a placebo in a double-blind study. There was an increase in lactase in the colon in the subjects who received lactose but not in the subjects who received sucrose. There was a decrease in colonic pH and also a decrease in breath H_2 excretion in those subjects who consumed lactose. However, the symptoms of lactose intolerance, i.e. pain, flatus, bloating and borborigmi, decreased for all subjects. It is important to recognise the possibility of a placebo effect in the studies dealing with "adoption" because, when volunteers attend the study for the first time in a unit, they are stressed and they have more symptoms than when they are seen for the second time.

10.9   Visceral Sensitivity

Visceral sensitivity varies between individuals; some people are very sensitive to lactose and the most sensitive suffer from irritable bowel syndrome; other people are very resistant. Visceral sensitivity is a very important parameter for lactose tolerance.

Prevalence of lactose intolerance in Finland is quite high and Vesa studied 366 people with self-reported lactose intolerance (20). The participants were examined to determine whether they had lactose malabsorption, or irritable bowel syndrome. Thirty-seven per cent of the whole population studied claimed to be lactose-intolerant; however, only half of them were lactose maldigestors. Thirty-two per cent had IBS.

10.10   Treatment of Lactose Intolerance

The first option in the treatment of lactose intolerance is to adopt a lactose-free diet. However, this decreases calcium consumption and the best solution is to work with the subject to try to decrease the quantity of lactose that is malabsorbed by means of low-lactose diets, milk-free diets, and diets that avoid only liquid lactose.

There are two different situations in which lactose malabsorption or intolerance occurs: short-term disease and chronic disease. Short-term diseases are usually severe - for example, acute gastroenteritis, characterised by diarrhoea. Radiotherapy also causes digestive problems and the easiest solution is to prescribe a short-term diet.

The treatment of infants with acute gastroenteritis used to be to avoid all milk products. A study by Brown *et al.* (21) investigating the effect of lactose-containing and lactose-free diets in children with acute gastroenteritis showed that, when children were dehydrated, treatment with a lactose-free formula resulted in a significant decrease in the risk of persistant diarrhoea. Therefore, prescription of a lactose-free diet is appropriate in the few children who have real dehydration. However, in children who do not have severe dehydration there is no beneficial effect of consuming a lactose-free diet (22).

Care should be taken in prescribing long-term diets - for example, in patients with short bowel, gastrectomy, or with hypolactasia, and those who think they have lactose intolerance. Excessively restrictive and *a priori* diets should be avoided, and the best advice is to try to follow a milk-free diet for a few days and evaluate symptoms.

In a study by Marteau *et al.* (22), malabsorption of milk was investigated in 14 patients with short bowel syndrome, with a bowel length of 60 cm, in whom half of the milk consumed is not digested. Many physicians would advise a lactose-free diet for these subjects. In this study, the subjects consumed 20 g of lactose, and a lactose-free diet in two different dosing periods. The diet containing 20 g lactose did not induce more symptoms and did not significantly increase faecal weight. The lactose dosing diet contained 20 g of lactose per day but less than 4 g in the form of milk; other lactose-containing foods included yoghurt and cheese.

10.11 Conclusions

Lactose malabsorption is a prevalent condition; however, lactose intolerance to doses found in the normal diet is rare. There is high variability of lactose malabsorption between patients. There is dose dependency of lactose malabsorption with regard to symptoms, depending on the nature of the lactose-containing products; solid products, or products with lactase are very well tolerated most of the time. The subjective nature of the symptoms of self-reported lactose intolerance should be taken into account, and, if a particular diet is prescribed to reduce symptoms of lactose malabsorption, calcium intake should be considered.

10.12 References

1. Vesa T.H., Marteau P., Korpela R. Lactose intolerance. *J. Am. Coll. Nutr.*, 2000, 19 (2 Suppl), 165S-175S.

2. Vesa T.H., Korpela R.A., Sahi T. Tolerance to small amounts of lactose in lactose maldigesters. *American Journal of Clinical Nutrition*, 1996, 64 (2), 197-201.

3. Buts J.P., Bernasconi P., Van Craynest M.P., Maldague P., De Meyer R. Response of human and rat small intestinal mucosa to oral administration of Saccharomyces boulardii. *Pediatric Research*, 1986, 20 (2), 192-6.

4. Lerebours E., N'Djitoyap Ndam C., Lavoine A., Hellot M.F., Antoine J.M., Colin R. Yogurt and fermented-then-pasteurized milk: effects of short-term and long-term ingestion on lactose absorption and mucosal lactase activity in lactase-deficient subjects. *American Journal of Clinical Nutrition*, 1989 May, 49, 5, 823-7.

5. Marteau P., Flourie B., Pochart P., Chastang C., Desjeux J.F., Rambaud J.C. Effect of the microbial lactase (EC 3.2.1.23) activity in yoghurt on the intestinal absorption of lactose: an in vivo study in lactase-deficient humans. *British Journal of Nutrition*, 1990, 64 (1), 71-9.

6. Arrigoni E., Marteau P., Briet F., Pochart P., Rambaud J.C., Messing B. Tolerance and absorption of lactose from milk and yogurt during short-bowel syndrome in humans. *American Journal of Clinical Nutrition*, 1994, 60 (6), 926-9.

7. Martini M.C., Lerebours E.C., Lin W.J., Harlander S.K., Berrada N.M., Antoine J.M., Savaiano D.A. Strains and species of lactic acid bacteria in fermented milks (yogurts): effect on in vivo lactose digestion. *American Journal of Clinical Nutrition*, 1991, 54 (6), 1041-6.

8.   Martini M.C., Kukielka D., Savaiano D.A. Lactose digestion from yogurt: influence of a meal and additional lactose. *American Journal of Clinical Nutrition*, 1991, May, 53, 5, 1253-8.

9.   Kotz C.M., Furne J.K., Savaiano D.A., Levitt M.D. Factors affecting the ability of a high beta-galactosidase yogurt to enhance lactose absorption. *Journal of Dairy Science*, 1994, 77, 12, 3538-44.

10. Martini M.C., Lerebours E.C., Lin W.J., Harlander S.K., Berrada N.M., Antoine J.M., Savaiano D.A. Strains and species of lactic acid bacteria in fermented milks (yogurts): effect on in vivo lactose digestion. *American Journal of Clinical Nutrition*, 1991, 54, 6, 1041-6.

11. Steidler L., Hans W., Schotte L., Neirynck S., Obermeier F., Falk W., Fiers W., Remaut E. Treatment of murine colitis by Lactococcus lactis secreting interleukin-10. *Science*, 2000, 289-5483, 1352-5.

12. Drouault S., Juste C., Marteau P., Renault P., Corthier G. Oral Treatment with Lactococcus lactis Expressing Staphylococcus hyicus Lipase Enhances Lipid Digestion in Pigs with Induced Pancreatic Insufficiency. *Appl. Environ. Microbiol.*, 2002, 68 (6), 3166-8.

13. Leichter J. Comparison of whole milk and skim milk with aqueous lactose solution in lactose tolerance testing. *American Journal of Clinical Nutrition*, 1973, 26, 4, 393-6.

14. Welsh J.D., Hall W.H. Gastric emptying of lactose and milk in subjects with lactose malabsorption. *American Journal of Digestive Disease*, 1977, 22, 12, 1060-3.

15. Lee C.M., Hardy C.M. Cocoa feeding and human lactose intolerance. *American Journal of Clinical Nutrition*, 1989, 49, 5, 840-4.

16. Nguyen K.N., Welsh J.D., Manion C.V., Ficken V.J. Effect of fiber on breath hydrogen response and symptoms after oral lactose in lactose malabsorbers. *American Journal of Clinical Nutrition*, 1982, 35, 6, 1347-51.

17. Martini M.C., Savaiano D.A. Reduced intolerance symptoms from lactose consumed during a meal. *American Journal of Clinical Nutrition*, 1988, 47, 1, 57-60.

18. Mahe S., Marteau P., Huncau J.F., Thuilier F., Tomé D. Intestinal nitrogen and electrolyte movements following fermented milk ingestion in human. *British Journal of Nutrition*, 1994, 71, 169-80.

19. Briet F., Pochart P., Marteau P., Flourie B., Arrigoni E., Rambaud J.C. Improved clinical tolerance to chronic lactose ingestion in subjects with lactose intolerance: a placebo effect? *Gut*, 1997, 41, 5, 632-5.

20. Vesa T.H., Seppo L.M., Marteau P.R., Sahi T., Korpela R. Role of irritable bowel syndrome in subjective lactose intolerance. *American Journal of Clinical Nutrition*, 1998, 67, 4,710-5.

21. Brown K.H., Peerson J.M., Fontaine O. Use of nonhuman milks in the dietary management of young children with acute diarrhea: a meta-analysis of clinical trials. *Pediatrics*, 1994, 93, 1,17-27.

22. Marteau P., Messing B., Arrigoni E., Briet F., Flourie B., Morin M.C., Rambaud J.C. Do patients with short-bowel syndrome need a lactose-free diet? *Nutrition*, 1997, 13, 1, 13-6.

11. COELIAC DISEASE: A CLINICAL UPDATE

Professor Peter Howdle

11.1   Introduction

Coeliac disease occurs in genetically susceptible individuals in whom there is an abnormality of the small intestinal mucosa made manifest by contact with the prolamin fractions of wheat (gluten and gliadin), rye (secalin) or barley (hordein). The majority of the symptoms that occur are a result of the mucosal abnormality characteristic of coeliac disease, and patients commonly suffer weight loss, bowel disturbance and lethargy. Coeliac patients are able to consume cereals such as maize, millet, rice.

11.2   Intestinal Morphology

The intestinal mucosa of an individual without coeliac disease is formed into finger-shaped villi. The surfaces of the villi are covered by cells called enterocytes which are used for absorption of nutrients from the gut, and the arrangement of villi in their finger-like form provides the gut with a large surface area to perform this function. In a patient with severe coeliac disease, the intestinal mucosa has no villi. There only remain the crypts, normally between villi, which become elongated in coeliac disease.

In diagnosing coeliac disease, because it is a gastrointestinal problem of the small intestine, one would expect patients to have gastrointestinal symptoms, and this has been very much the case in the majority of patients although recently there has been a change in the clinical presentation observed in many patients.

11.3   Diagnosis of Coeliac Disease

In 1992 Howdle *et al.* (unpublished observations) examined the symptoms of 49 new cases of coeliac disease that were diagnosed in 1990: 70% of patients had had diarrhoea; almost 70% had had weight loss, which is associated with a disease of malabsorption such as coeliac disease; and 76% of patients had been feeling weak and lethargic and lacking in energy.

Review of the symptoms associated with coeliac disease over the last four decades showed that, in the 1950s, 90% of patients were suffering the same symptoms of weakness, weight loss, and diarrhoea. These symptoms became slightly less common through the 1980s and 1990s, but were still relatively common. In the 1970s, 1980s and 1990s, children with coeliac disease would appear to be unhappy, with a lack of appetite and failure to gain weight, coupled with irritability.

However, in recent years the clinical presentation has been changing. In Bologna, Corazza et al. (1) followed the case histories of patients diagnosed as having coeliac disease, and showed that, towards the end of the 1980s the number of patients presenting with the classic symptoms of diarrhoea and weight loss reduced, and there were almost twice as many patients with minor symptoms. The majority of these, in fact, presented with anaemia. Over the last 10 years, this trend has been confirmed (2,3).

The study by Hin et al. (4) examined the symptoms of 1,000 patients who had presented to their GP over the course of one year with the symptoms listed in Table 11.I. The number of patients with those symptoms and those diagnosed with coeliac disease are also shown in Table 11.I.

The patients were screened for coeliac disease and, of the 1,000 patients examined, 30 had a positive blood test for coeliac disease; all these patients had an abnormal biopsy of the duodenum. Six patients whose only complaint was feeling tired all the time, were found to have coeliac disease, and this was an important finding in a general practice setting. It was concluded that the 30 patients were new coeliac patients out of the 1,000 that had presented with symptoms that might have suggested coeliac disease. The incidence of coeliac disease in the study population was therefore 3%; however, investigators estimate that the working prevalence of coeliac disease in the population is 1 in 300 (i.e. 0.3%). As a result of this research, it has been suggested that, in 12% of people with anaemia, coeliac disease may be the cause. None of the patients in this study with symptoms of irritable bowel syndrome had coeliac disease, although another study (5) has shown that between 5 and 10% of people with irritable bowel might have coeliac disease.

TABLE 11.I
Symptoms of coeliac disease

Symptom	Number of people with symptoms	Number of people identified with coeliac disease
Irritable bowel syndrome	132	0
Anaemia	126	15
Family history of coeliac disease	28	2
Malabsorption/diarrhoea	93	5
Fatigue	329	6
Endocrine complaint (thyroid, diabetes)	157	1
Paediatric problem (weight loss, failure to thrive)	36	0
Other (abnormal blood test, epilepsy, infertility, arthralgia)	99	1

The key finding of this research was that general practitioners currently see many patients who may have undiagnosed coeliac disease, who present with symptoms not immediately associated with the condition. A prevalence figure of 1% for coeliac disease is now quoted for the general population (6).

Data from Coeliac UK regarding new members that joined in 1996 shows that, in children, the numbers of new patients have not increased; however, approximately twice as many adult patients joined the society as 10 years previously, and the mean age, or peak age, of those joining has moved from between 30 to 35 up to approximately 55 or 60. One hundred and twenty new patients over the age of 75 joined Coeliac UK in 1996. It is assumed that all those people joining Coeliac UK are newly diagnosed coeliac sufferers.

This finding confirms that there are many more patients being diagnosed than there were previously. Many of these are patients with more minor symptoms and will probably be older having perhaps been mildly unwell for many years. It is only when the possibility of coeliac disease is thoroughly investigated that it is diagnosed.

11.4   Screening Tests

Two antibody screening tests are used to detect coeliac disease: the anti-endomysial antibody test (EMA) and the anti-tissue transglutaminase antibody test. The EMA is an immunofluorescence test which looks for antibodies against a reticulum tissue – in monkey oesophagus or human umbilical cord. This is a specific test performed using serum from a coeliac sufferer and is approximately 95% accurate for diagnosing coeliac disease. The antibodies against endomysial tissue are acting against an enzyme called tissue transglutaminase. An Enzyme-linked immunosorbent assay (ELISA) test against this enzyme produces good results, similar to the EMA test.

These types of screening test have been used in normal populations, i.e. healthy populations of people with no symptoms of coeliac disease. Two studies conducted in school children in Italy suggested coeliac disease was present in 1 in 250 in children (7,8). Johnson *et al.* (9) suggested coeliac disease in 1 in 140 adults tested, and Volta's study in Italy reported a similar prevalence figure in adults (10). The study by West *et al.* found that 1 in 77 people tested proved positive for coeliac disease using these screening tests (6). This gives some indication of the amount of undiagnosed coeliac disease most probably present in the general population.

11.5   Diseases Associated with Coeliac Disease

It is well known that certain diseases have an association with coeliac disease; for example, among people with insulin-dependent diabetes, a significant number will develop coeliac disease, probably related to the similar genetic inheritance. Of people with chronic thyroid disease, or primary biliary cirrhosis, it is expected that a significant number will also suffer from coeliac disease.

Osteoporosis is obviously a big problem in the community, particularly post-menopausal osteoporosis, and it is often thought to be idiopathic. However, if people with osteoporosis, who seem to develop it for no apparent reason, are examined, a significant number will have coeliac disease. Nuti *et al.* recently suggested that coeliac disease was the cause of osteoporosis in 9.4% of patients with osteoporosis (11).

Coeliac sufferers also tend to have problems with fertility (12).

A significant number of people with neurological symptoms such as ataxia, the cause of which may be very difficult to diagnose, have some problem with gluten or gliadin intolerance and may have coeliac disease. There is a lot of research going on in this area currently (13).

There is a strong association of coeliac disease with Down's Syndrome. Coeliac disease does run in families; approximately 10% of first degree relatives also have the disease. Therefore, if patients in this group are investigated, there could be several undiagnosed cases of coeliac disease that have arisen through genetic inheritance.

Figure 11.1 shows the Coeliac Iceberg (2). If all people with the right genetic susceptibility to coeliac disease were examined, some people would have symptoms and be diagnosed. Some people would have had no symptoms, but would have an abnormal biopsy. These patients suffer from what is described as silent coeliac disease. There will be some patients who would have a normal or near-normal biopsy, and it is believed that they are likely to develop coeliac disease (latent coeliac disease). There are also healthy individuals who have the potential to have coeliac disease through genetic inheritance but do not seem to develop the disease.

11.6 Interpretation of Intestinal Biopsy

A problem in diagnosing coeliac disease is in determining the definition of an abnormal biopsy. In a normal histological section of the small intestine, the finger-like villous shape is clearly defined, with enterocytes covering the villous surface. Absorption takes place across these enterocytes in normal digestion and absorption. In a person suffering from coeliac disease there are no villi in the intestine; the mucosa is flat, and the crypts become elongated and full of inflammatory cells called lymphocytes. These are the classical histological features of untreated coeliac disease.

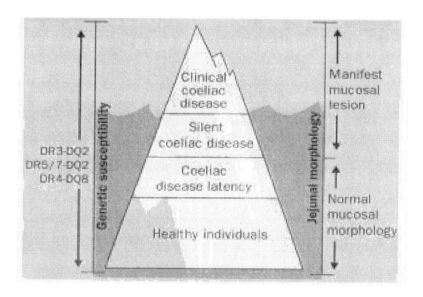

Fig. 11.1. Coeliac disease iceberg and spectrum of gluten sensitivity.
Reprinted with permission from Elsevier Science from Maki M., Collin P. Coeliac
disease. Lancet, *1997, 349, 1755-59.*

After taking biopsies from the upper small intestine (duodenum) in order to make a diagnosis, if the typical severe features of coeliac disease are present it is straightforward - they are easily recognised; however, more patients are now being found who have only minimal changes to the gut mucosa and it may be necessary to repeat the biopsy test on more than one occasion to find if it really is abnormal. Observing the effect of changing diet, such as omitting rye, wheat and barley, on gut morphology can be useful.

The problems of diagnosing coeliac disease using biopsy is demonstrated by the following case study. A biopsy from a patient who presented with anaemia 10 or 12 years ago appeared to be only slightly abnormal. The patient was generally healthy and leading a normal life, but had been found to be anaemic for no apparent reason. There were villi present on the gut mucosa, and the crypts were not very deep. There were not a lot of inflammatory cells, although there were a lot of lymphocytes amongst the enterocytes; these are intra-epithelial lymphocytes. The patient was advised to consume plenty of gluten and wheat, and a biopsy was taken to determine whether the consumption of gluten resulted in changes to the mucosa that would be expected in coeliac disease. The biopsy that was taken a few months later showed tall thin villi. In coeliac disease one would expect the villi to become shortened; however, there were a lot of lymphocytes still present. The patient ceased to attend the clinic but was treated for anaemia with oral iron. Eight years later the patient returned to the clinic with anaemia and the biopsy showed an absence of villi, and the presence of deep crypts and inflammatory cells, indicating typical untreated coeliac disease. This was an interesting case because it demonstrated that the disease evolved over a period of time and it is not common to be able to observe such a progression.

Figure 11.2 demonstrates the sort of abnormalities that can occur in biopsies in coeliac patients. At the pre-infiltrative point, the villi are normal (type 0), and the first change that can be observed is that the villi are still normal but lymphocytes infiltrate (type 1). In the hyperplastic (type 2) state, the immune reaction that is believed to be the cause of the disease begins to be apparent, and the lymphocytic infiltrate increases. In this hyperplastic stage, the villi still appear to have a normal shape, but the crypts are deeper and more lymphocytes infiltrate. Finally, in the normal circumstances of advanced coeliac disease, the intestinal mucosa takes on the flat destructive form (type 3), which is the classical, typical type of biopsy appearance seen in new coeliac patients. If patients follow a gluten-free diet – omitting the cereals related to coeliac disease from their diet – the intestinal mucosa will recover its shape and return towards normal. However, if patients continue to consume gluten, the intestinal mucosa will remain abnormal and rarely may become atrophic (type 4). The mucosa appears to be able to heal itself once gluten is omitted from the diet. However, there is a huge variation in sensitivity to gluten and in some patients, the mucosa can recover and heal from exposure to gluten quickly even if small amounts of gluten are inadvertently in the diet, but in others the mucosa will recover much more slowly, as even a trace of gluten can prevent proper mucosal recovery.

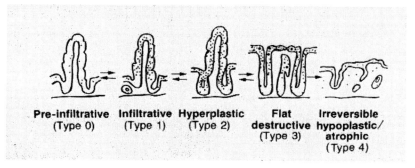

Pre-infiltrative   Infiltrative   Hyperplastic   Flat   Irreversible
   (Type 0)   (Type 1)   (Type 2)   destructive hypoplastic/
                                              (Type 3)   atrophic
                                                           (Type 4)

Fig. 11.2. Gut mucosa changes in development of coeliac disease.
Reproduced with kind permission from Marsh M.N. Gluten, major
histocompatibility complex, and the small intestine. A molecular and
immunobiologic approach to the spectrum of gluten sensitivity ('coeliac sprue').
Gastroenterology, 1992, 102 (1), 330-54.

A small number of patients are diagnosed with irreversible hypoplastic gut mucosa. They have usually been undiagnosed and have suffered from the disease for a long time. They have severe digestive problems, and the gut never recovers. This is at the very severe end of the spectrum but it does appear in a small number of patients.

Approximately 90% of patients will present with flat destructive type 3 mucosa and, after following a gluten-free diet, the biopsy will show mucosa that is almost normal. Some patients present with type 3 flat destructive mucosa and do not improve much; however, the vast majority of patients fit into a relatively straightforward diagnostic pattern.

11.7   Pathogenesis of Coeliac Disease

Figure 11.3 demonstrates the pathogenesis of coeliac disease. The current hypothesis is that gliadin, the protein from wheat, and some of the proteins from rye and barley, are broken down in the intestinal lumen and the peptide fragments of gliadin are absorbed across the enterocyte membrane into the lamina propria. These gliadin peptides are deaminated by tissue transglutaminase, and the processed peptide is taken up by an antigen-presenting cell. This is an immune cell that has the right histocompatibility antigen to present the processed gliadin to the effector lymphocytes, which produce an inflammatory response in the form of cytokines, and/or antibodies thereby affecting the mucosa. This immune response damages the mucosa and causes the pathological changes described above. Why this occurs in some patients with the right genetic inheritance, and not in others, is still an open question.

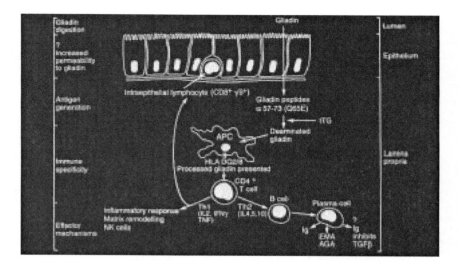

Fig. 11.3. Role of gliadin, tissue transglutaminase (tTG) and histocompatibility antigens in generating the immune response in coeliac disease. Potential effector mechanisms that may cause mucosal damage are illustrated.
Reproduced with kind permission from Jennings J.S.R., Howdle P.D. Coeliac disease. Current Opinion in Gastroenterology, 2001, 17, 118-26.

11.8   Complications of Coeliac Disease

The major complications of coeliac disease are rare. Severe ulceration and narrowing of the small intestine can occur but there are no good prevalence data on this complication of the disease and most clinicians have generally seen very few such cases. Most patients who stick to a gluten-free diet should avoid this complication.

11.8.1 Malignancy of the small intestine

Lymphoma and adenocarcinoma of the small intestine are very rare tumours in the normal population. Of the patients that have these tumours, a significant number do have coeliac disease. However, a national study into the link between coeliac disease and these small intestinal tumours has recently been completed (16). The study found that, if someone has coeliac disease, the chances of actually getting a lymphoma are extremely small, and this reduces to virtually zero if they stick to a strict gluten-free diet. Therefore, lymphoma is a complication of coeliac disease, but is perhaps more rare than was previously thought.

11.8.2 Osteoporosis

The definition of osteoporosis agreed by the World Health Organization is that it is characterised by low bone mass and micro-architectural deterioration. Studies that have reviewed osteoporosis have found that there is a high incidence of osteoporosis and thinner bones in patients with coeliac disease (17,18).

In a study conducted over the course of a year, it was shown that, if patients who had not previously been treated for coeliac disease consumed a gluten-free diet, the bone density in the spine improved. Therefore, treating coeliac disease does in fact improve the bones of these patients. There are some possible causes of osteoporosis in coeliac disease. Patients can be malnourished, they take less calcium and do not perhaps absorb it as efficiently as normal.

11.9   Treatment of Coeliac Disease

The best treatment for coeliac disease is a strict gluten-free diet. There has been a question about the suitability of oats in the diet and there have been two initial studies that suggested that oats were safe for coeliacs to eat (19,20). Current thinking is that coeliac sufferers may consume oats without suffering adverse effects. However, there have been some caveats; some of the early studies on oat consumption were criticised because only small doses of oats were given to patients, and the long-term effects of consumption of oats were not taken into consideration. On following up the early studies however it would appear that consumption of oats is safe. The only problem raised is whether oats that are made commercially, are made in mills and manufacturing conditions that are free from other cereals.

Iron and folic acid supplementation is also advised. A re-biopsy may be necessary, particularly if patients have not responded well to treatment.

11.10   References

1.   Corazza G.R., Frisoni M., Treggiari E.A., *et al.* Subclinical celiac sprue. Increasing occurrence and clues to its diagnosis. *Journal of Clinical Gastroenterology*, 1993, 16, 16-21.

2.   Maki M., Collin P. Coeliac disease. *Lancet*, 1997, 349, 1755-59.

3.   Feighery C. Coeliac disease. *British Medical Journal*, 1999, 319, 236-9.

4.   Hin H., Bird G., Fisher P., Mahy N., Jewell D. Coeliac disease in primary care: case finding study. *British Medical Journal*, 1999, 16, 318, 7177, 164-7.

5.   Sanders D.S., Carter M.J., Hurlstone D.D., *et al.* Association of adult coeliac disease with irritable bowel syndrome: a case-control study in patients fulfilling ROME II criteria referred to secondary care. *Lancet*, 2001, 358, 1504-8.

6.   West J., Lloyd, C.A., Reader R., Hill P.G., Holmes G.K.T., Khaw K.T. Prevalence of misdiagnosed coeliac disease in the general population of England. *Gut*, 2001, 48, A63.

7.   Mazzetti di Pietralata M., Giorgetti G.M., Gregori M., De Simone M., Leonardi C., Barletta P.A., Ricciardi M.M., Sandri G. Subclinical coeliac disease. *Italian Journal of Gastroenterology*, 1992, 24, 6, 352-4.

8. Catassi C., Ratsch I., Fabiani *et al*. Coeliac disease in the year 2000; exploring the iceberg. *Lancet*, 1994, 343, 200-3.

9. Johnson S.D., Watson R.G.P., McMillan *et al*. Prevalence of coeliac disease in Northern Ireland. *Lancet*, 1997, 350, 1370.

10. Volta U., Granito A., De Franceschi L., Petrolini N., Bianchi F.B. Anti-tissue transglutaminase antibodies as predictors of silent coeliac disease in patients with hypertransaminasaemia of unknown origin. *Dig Liver Dis*, 2001, 33, 5, 420-5.

11. Nuti R., Martini G., Valenti R., Giovani S., Salvadori S., Avanzati A. Prevalence of undiagnosed coeliac syndrome in osteoporotic women. *Journal of International Medicine*, 2001, 250, 361-6.

12. Sher K.S., Jayanthi V., Probert C.S., Stewart C.R., Mayberry J.F. Infertility, obstetric and gynaecological problems in coeliac sprue. *Digestive Diseases*, 1994, 12, 186-90.

13. Hadjivassiliou M., Gibson A., Davies-Jones G.A.B., Lobo A.J., Stephenson T.J., Milford-Ward A. Does cryptic gluten sensitivity play a part in neurological illness? *Lancet*, 1996, 547, 369-71.

14. Marsh M.N. Gluten, major histocompatibility complex, and the small intestine. A molecular and immunobiologic approach to the spectrum of gluten sensitivity ('coeliac sprue'). *Gastroenterology*, 1992, 102 (1), 330-54.

15. Jennings J.S.R., Howdle P.D. Coeliac disease. *Current Opinion in Gastroenterology*, 2001, 17, 118-26.

16. Howdle *et al*. (in preparation).

17. McFarlane X.A., Bhalla A.K., Reeves D.E., Morgan L.M., Robertson D.A. Osteoporosis in treated adult coeliac disease. *Gut*, 1995, 36 (5), 710-4.

18. Walters J.R., Banks L.M., Butcher G.P., Fowler C.R. Detection of low bone mineral density by dual energy x-ray absorptiometry in unsuspected suboptimally treated coeliac disease. *Gut*, 1995, 37 (2), 220-4.

19. Janatuinen E.K., Pikkarainen P.H., Kemppainen T.A., Kosma V.M., Jarvinen R.M., Uusitupa M.I., Julkunen R.J. A comparison of diets with and without oats in adults with coeliac disease. *New England Journal of Medicine*, 1995, 19, 333 (16), 1033-7.

20. Srinivasan U., Leonard N., Jones E., Kasarda D.D., Weir D.G., O'Farrelly C., Feighery C. Absence of oats toxicity in adult coeliac disease. *British Medical Journal*, 1996, 23, 313, 7068, 1300-1.

12. COELIAC DISEASE: A CONSUMER PERSPECTIVE

Mr Robert Palgrave

12.1   Introduction

The treatment for coeliac disease is a gluten-free diet. Plini lived in ancient Roman times and wrote a book called "Natural History", which was translated into English in 1959. Part of it talks about remedies from women. In the book he says: "As to the use of woman's milk it is agreed that it is the sweetest and most delicate of all, very useful in long fevers, and coeliac disease, especially the milk of a woman who has already weaned her own child." So even back in those days it was known that a gluten-free diet was the treatment for coeliac disease.

Food is frequently mentioned in other literature and the quotes below give some idea of the importance of wheat-containing foods in the staple diet, and what a person suffering from coeliac disease has to deal with in living with a gluten-free diet for maybe 60 or 70 years.

References to food in literature, old and new
"Go thy way, eat thy bread with joy"
Ecclesiaastes 9.7

"Give us this day our daily bread"
The Lord's Prayer

"Part of the secret of success in life is to eat what you like and let the food fight it out on the inside"
Mark Twain

The point about religion and bread is quite serious for coeliacs; communion wafers are made from wheat. The Pope recently decreed that communion wafers made from gluten-free wheat are not sacrament and do not represent the body of Christ, which puts coeliac Roman Catholics in something of a dilemma.

An important point to make is that there is food available that is suitable for a coeliac sufferer; however the coeliac has to be informed about the restrictions in diet that are associated with coeliac disease, and take responsibility for managing his or her own diet. Food is not the only problem; coeliacs can drink alcohol, wine, cider, and spirits, but they cannot drink beer, which contains gluten. Advances in technology are being made towards development of beer that is suitable for coeliacs.

For many people there is a stigma associated with having a disease. For example, there are 650 Members of Parliament at the moment in Westminster and not one of those is a member of the Coeliac Society, now called Coeliac UK. The prevalence of coeliac disease in the general population is as high as one in a hundred, so it is surprising that Coeliac UK has never had an MP as a member. Celebrities rarely admit that they have got coeliac disease.

12.2 Positive Aspects of Coeliac Disease

Most people understandably focus on the negative aspects or possible negative aspects of being coeliac; however, there are some aspects of the disease that could be construed as beneficial. The nature of the disease means that food is not as efficiently digested, which means that coeliacs are less at risk from obesity; one of the common characteristics of coeliacs is that they are generally lean. Coeliacs also suffer less from heart and stroke problems. It is known that rats live longer on a low-calorie diet, so perhaps coeliacs will live longer because they have less food going through their system.

12.3 Negative Aspects of Coeliac Disease

Negative aspects of coeliac disease include diarrhoea, which can seriously disrupt a person's normal working life. Lower energy and sexual performance are also common features, as is osteoporosis and lower reproductive success; coeliac parents take longer to conceive and have more problems with miscarriages. Coeliacs also suffer from dental and oral health problems, the causes of which are not readily understood, and include problems with teeth and gums, and mouth ulcers. Adoption of a gluten-free diet once coeliac disease has been diagnosed often resolves these problems. Coeliacs also suffer from skin and hair problems such as alopecia and psoriasis, which also resolve on adoption of a gluten-free diet.

A recent study published from the University of Sheffield has shown that people who have coeliac disease are more likely to have persistent unexplained headaches. These unexplained headaches may cease with a gluten-free diet, but this is not always the case.

Coeliacs may have a higher risk of malignant tumours than the general population.

Children of coeliacs have a lower birth weight than average, and children are often born pre-term. A study conducted in Sweden has reported that pre-term birth affects infants of coeliacs if either the father or the mother suffers from the condition.

12.4   Diagnosis

People present with a wide range of symptoms, which might include only anaemia and constant fatigue. Following diagnosis and adoption of a gluten-free diet, some people realise that the disease has compromised their quality of life for as long as 20 or 30 years, and a gluten-free diet dramatically improves their quality of life.

People have very different reactions to what the treatment is and the practical problems associated with living with the disease. Some newly diagnosed coeliacs are relieved that the disease does not require surgery or drugs to treat it. Others are dismayed that a wide range of foods that they ate regularly will be eliminated from their diet. In some cases, the reaction is extreme and can involve social exclusion, because the coeliac has trouble adjusting to his/her new diet.

Many coeliacs ask Coeliac UK whether there is a cure or a vaccine for coeliac disease. Unfortunately, there is no cure, and the only means of living with the disease is by adopting a gluten-free diet. People are often concerned about the effects of their previous diet on their body before coeliac disease was diagnosed. Questions include: Is there permanent damage as a result of having eaten gluten for 15 or 20 years? Am I very sensitive to gluten? Can I take the odd risk? If the symptoms of coeliac disease are not very pronounced it might be tempting to carry on eating small quantities of gluten every day, without realising that this will contribute to problems in the future.

The most important question is: "Where can I get reliable information about the gluten content of the food I am buying in the shops or the food I am being served in the restaurant?" What Coeliac UK, previously the Coeliac Society, has done for many years is to produce an annual food list that is a collection of information from food manufacturers who have made a self-declaration about the gluten content of food. The list is updated on a monthly basis.

12.5 Genetic Aspects of Coeliac Disease

Many coeliacs ask whether they have passed the disease on to their children, whether their children should be tested and how can they be prevented from getting coeliac disease as well. It is known that there is a genetic susceptibility for coeliac disease, and then perhaps some event, some environmental factor, comes into play that causes the condition to develop.

12.6 Managing a Gluten-free Diet

Coeliac disease is expensive in that gluten-free foods are more expensive than wheat-containing products and not all coeliacs are in the position to be able to afford specialist foods. After diagnosis, some of the difficulties that coeliacs face include not knowing their limits for consuming gluten. The test to determine whether the gut is being damaged by diet is a biopsy, which obviously cannot be performed regularly, so individuals must manage their diet themselves.

If symptoms are not severe, it is easier to pretend that you have not eaten food containing gluten than to insist, in certain social situations, on food that is gluten-free, or to read the label on every food consumed, and to accept the word of a waiter in a restaurant who tells you that that food that he has just put in front of you is gluten-free.

Restaurants are a typical environment where consuming food that has been prepared by somebody else carries a risk that it contains gluten. It is easy to identify foods that are not suitable - for example a pie or pasta; however, the advice of a waiter or chef must be relied upon for whether some dishes contain a gluten-containing grain. Very rarely do menus state suitability of individual dishes for coeliacs.

General advice to coeliac sufferers is to place less reliance on convenience foods or processed foods and snacks. Products that replace those that typically contain wheat but are not available on prescription will be a lot more expensive than the regular versions.

Some specific challenges to sticking to the diet are encountered by the "microwave generation" - people who are under 40 and have grown up opening packets, putting them in the microwave and eating the results. People in this section of the population have busy lives, and want flexibility and spontaneity. They are not particularly interested in cooking. This is a huge generalisation, of course, but this socio-economic group is generally more interested in using convenience foods and eating out. They might deny having the condition if they do not have any real symptoms.

The generation of people who are over 50 years old may have to change the dietary habits of a lifetime. They may eat bread every day and will have to adapt to that change, and learn to read product labels. Hospitals and nursing homes can also be a problem for people with coeliac disease because the catering there is not necessarily tailored to providing for special diets.

A child being coeliac can affect school meals and school trips. There is a lot of advertising pressure on people to eat particular foods, which may have gluten in them, and children often want to try foods that they have seen advertised, or their friends are eating. Well-meaning adults might hand out biscuits or chocolate bars that have got wafers in them. Of course, the pressure is on parents to try to manage the diet for the child.

The biggest problem is in the adolescent age group. There is a stigma for an adolescent to be seen to be different from his/her friends. Coeliac disease may be seen as a problem for the future, i.e. when they are 22, or 23; then they will start looking after themselves. They like fast food and have active social lives. It is easy to imagine the problem for a teenager of 18 or 19 - you cannot drink beer or eat pizzas; you cannot go out for fish and chips or a Chinese take-away because there is possibly wheat flour in some of the sauces, and it is unlikely that you could safely eat every dish in an Indian restaurant.

Dietary management of a coeliac involves consumption of naturally gluten-free foods. To achieve this, grain-based starches are replaced with foods such as potato, rice, maize, buckwheat, etc. Beer is not allowed and fruit and vegetables should be increased, which is a general dietary recommendation in itself. A gluten-free diet becomes more feasible by spending more time in the kitchen preparing food from raw ingredients. This involves understanding how meals are prepared and what goes into food. It is easy to bake your own bread, which is very cost-effective, and results from gluten-free flours are very good.

Cross-contamination with gluten in the kitchen for a coeliac is very important, so it is vital to be aware of possible areas of contamination, and make appropriate arrangements, such as having a toaster for wheat bread and a separate toaster for gluten-free bread. It is also important to tell friends, family and your employer about coeliac disease so that they can support you.

Special diets can be very bland; however, gluten-free food has improved dramatically in recent years.

It is also very difficult for some people to understand what "gluten-free" really means. Oats are now on the acceptable list for coeliacs, but there are approximately 20 different food names that people have to be aware of to avoid gluten. There are also problems with explaining "gluten-free" to shop assistants, waiters, friends, and perhaps your employer if you eat in the staff canteen. Getting the message across is the responsibility of the coeliac himself or herself.

A coeliac must accept that coeliac disease is a life-long condition. It affects impulse eating and drinking, from what drink to have in a pub, to whether you can have a bag of peanuts - you must look at the labelling on the bag of peanuts to see whether it has got any coating on it. Crisps may also contain gluten; plain crisps are probably gluten-free, but flavoured crisps are almost certainly not. It is difficult in social situations to be controlled by diet, and the reaction is to be less vigilant in finding out exactly what a food product contains.

12.7 Labelling

Knowing how to read and interpret food labels is very important. The rules governing what is included in a food label are complex. The 25% exemption rule, whereby, if compound ingredient makes up less than 25% of the product, the components making up the compound ingredient do not need to be listed on the ingredients label, with the exception of functional ingredients, means that a product may contain gluten, but this is not obvious on the product label. If gluten is used as a processing aid, it does not have to be declared on the label, and again will not be obvious from the product label. Coeliac UK aims to encourage manufacturers to self-declare whether or not gluten is in a product, even though it may not be declared on the label.

12.8 Foods Specially for Coeliacs

In the UK, people who are medically diagnosed with coeliac disease can get gluten-free food on prescription. The list of gluten-free foods on prescription does not include everything that is manufactured as a speciality gluten-free food. The principle behind selecting the foods on the list is that they are staple foods such as breads, pastas, plain biscuits, flours, and mixes for making your own bread.

That list of recommended products is common across the whole country; however, individual GPs will interpret that list themselves. Some people will be allowed a free choice from anything on that list, and other people are restricted to very plain foods, and in some cases quite low quantities. The access to foods on the prescription list depends very much on the location of the GP surgery and the budget available.

Coeliac disease has been classed by the general media together with other unspecific food intolerances that do not necessarily have a clear diagnosis. This situation can cause scepticism, which is counterproductive to people that have been properly diagnosed, and creates the impression that a gluten-free diet is a lifestyle choice rather than a medical necessity.

12.9 Other Implications of Coeliac Disease

A coeliac in the family requires support from parents, partners and relatives. Travelling on holiday and on business is also an area of great concern to coeliacs. Coeliac UK deals with a lot of enquiries about airline food and there is always concern over whether a gluten-free meal that has been booked on a flight will turn up. To counter this, coeliacs have survival tactics; they take food with them in their carry-on luggage. However, when travelling on business, this can be extremely difficult if eating occasions arise without the time for any forward planning.

Suffering from coeliac disease will also affect selection for some occupations - for example, the Armed Forces in the UK either will not accept or are very reluctant to accept coeliacs. In America, if you develop coeliac disease and you are in the US Army you will be dismissed. However, the British Army has recently approached the Vegetarian Society to collaborate on developing vegetarian ration packs but still finds the dietary requirements of coeliacs prohibitive.

Coeliacs also have to pay higher insurance premiums, including life and travel insurance.

12.10 The Way Forward

Ways in which the food industry can help people with coeliac disease are to re-evaluate the formulation of products, for example the use of wheat starch, which puts foods containing it off limits to coeliacs and to people who have wheat intolerance or wheat allergy.

An improvement in food labelling is also desperately needed, and is the case for many different allergies. Coeliac UK is currently striving to introduce a standard where there will be a symbol on food that indicates "gluten-free". The foodservice industry is also a big challenge in that thousands or hundreds of thousands of people in this sector need to be educated in allergies and the safe methods of food preparation with respect to contamination by allergens, including gluten.

13. FOOD ALLERGY AND INTOLERANCE: WHAT CHALLENGES DO THEY POSE TO THE FOOD INDUSTRY?

Fiona Angus

13.1 Introduction

This chapter covers some of the practical challenges that food allergy and intolerance present to the food industry, and some of the measures that can be taken to meet those challenges.

The main challenges that the food industry faces in relation to food allergens occur as a result of the potential severity of the reactions to many allergens such as peanut and tree nuts. The main areas of concern are products that contain an allergenic ingredient that may not be declared on the label, products that become contaminated through inadequate company controls, and products that become contaminated through the nature of the product or the process.

The consequences of these situations are the severe and sometimes fatal reactions to food, which have been widely reported in the media in recent years. Most of the fatal incidents occur in restaurants and catering establishments, but there have been some cases relating to consumption of manufactured foods.

The most likely scenario is the receipt of a customer complaint, which is often the case where a customer has had either a mild or a severe reaction. However, there is increasing risk of lawsuits against food and drink companies from food allergen-related incidents and associated damage to the company name.

13.2 Case Histories

Cases of fatal reactions that have occurred in the UK, include a boy who had been allergic to milk for several years, and ate some crisps without looking at the label. Milk powder was used in the coating on the crisps, and was included in the ingredients labelling, but the child did not read the label and did not therefore identify the presence of milk powder in the product. He had an anaphylactic reaction; there was a delay in administration of adrenaline and he died.

Another example of an incident involving an anaphylactic reaction to packaged foods was a case of a 20-year-old woman who had had mild nut allergy from the age of 2 but had never had severe reactions. She bit into a chocolate that had a hazelnut in it and, as she was not carrying adrenaline, she went to a nearby chemist's shop for help, but the staff would not administer adrenaline without a prescription; there was a delay in administering the adrenaline and she died. Since this case, the Anaphylaxis Campaign has done quite a lot of work with pharmacists to try and resolve this issue.

A case of an allergic reaction that was not fatal involved a girl who reacted to milk in a lolly. The milk was not declared on the label and the company was fined £1,500, which, while not being a large amount, does show that prosecutions are being made, and in the future settlements for prosecutions of this kind will probably be much larger.

13.3 Common Allergens

There are many hundreds of ingredients that can cause food allergy and intolerance; however the "big eight" – peanuts, tree nuts, egg, fish, crustacea, wheat, milk and soya - account for 90% of the problem. There will always be consumers who suffer a reaction to an ingredient that is not included in this list but food manufacturers should concentrate on managing the ingredients listed above.

13.4 Minimising Risk

There are several areas that are important to minimise risk: groundwork; risk assessment; product development strategies, including operational considerations; packaging and labelling; recall policy; and it is also important to review and refine the company allergen policy as the product range changes.

13.4.1 Groundwork

In preparing an allergen policy, it is important to brief management on its importance. There is currently much greater awareness of the problem of allergens than in the past, but there are still some smaller companies that are not adequately addressing this issue. If you don't have the expertise in-house, a useful means of updating management on developments is to invite an expert on allergy to give a presentation.

13.4.2 Risk assessment

A risk assessment team will prioritise the biggest risk areas through auditing and assessment of Hazard Analysis and Critical Control Points (HACCP). This should be done in combination with the instigation of an integrated allergen management programme.

13.4.3 Product development

Consideration should be given to whether allergenic ingredients are needed in a recipe at all. In the case of the boy that died as a result of eating crisps that contained milk flavouring, the company in question has now taken milk out of the flavouring. Clearly, there will continue to be allergens present in products, but in some cases they will not provide any functional or organoleptic advantages and could be removed from the formulation.

When a product is reformulated, an allergen should be added only if it is essential to the improvement of taste or texture of the product. Particular attention should be paid to the components of flavourings and processing aids because there could be trace allergens present that might not be obvious.

Emerging allergens should also be monitored. Lupin flour is a good example of a new food whose use is becoming more mainstream - some companies have replaced soya flour with lupin flour as a result of GM concerns; however, individuals with peanut allergy have been shown to cross-react to lupin.

13.4.4 Operational considerations

There are many operational considerations that need to be taken into account: a HACCP audit should be done and, if a particular company does not have the expertise to do this, information on HACCP is widely available.

Staff should be trained to be aware of the gravity of food allergy. This needs to be an ongoing process to avoid complacency.

Dedicated factories, lines, personnel and equipment for a particular allergen are the ideal, and many companies have gone a long way towards trying to achieve that. Where that is not possible, there are measures that can be taken, including adding the allergenic ingredients later on in the process, so that the whole line is not contaminated, and a shorter length of the production line needs to be thoroughly cleaned. Scheduling longer runs of product containing allergenic ingredients, and producing products containing particularly allergenic ingredients at the end of the week, near a major clean-down, are also advisable actions. The effectiveness of cleaning procedures should also be evaluated, and there are a number of kits available for this. Leatherhead Food RA runs an allergy testing service, and conducts regular clean-up checks for companies.

Care should be taken with the re-use of frying oils because traces of protein from one food that has been fried can be transferred into the oil and subsequently into the next food that is fried. Companies also need to have a re-work policy; use of re-work should be very carefully monitored to ensure that there is no potential for cross-contamination. The policy should be to use re-work only for the product from which it came. Re-work should be well labelled and segregated from other re-work. One effective method is to use the re-work within the same manufacturing run.

13.4.5 Suppliers

It is important to have allergen-related information from suppliers, particularly in relation to the top eight and with particular attention in relation to oils, i.e. whether cold pressed oils are used rather than refined oils. It is also important to audit suppliers regularly.

Co-packers should also be carefully audited. This is illustrated by a case where a co-packer was bottling peanut butter and marmalade on the same line. A consumer reacted to a fragment of peanut that was found in the marmalade, the third day into production. The company selling the marmalade did not know that peanut butter was being packed on the same line as the marmalade.

There is a huge number of packaging and labelling initiatives. A policy towards labelling the presence of the top eight is currently a voluntary practice but it is going to become mandatory over the next few years.

Some UK food retailers currently use "contains" labelling for the top eight allergens, and this is useful for alerting consumers to the presence of particular allergens.

13.5   Free-from Information

"Free from" information is available from Leatherhead Food RA via the Food Intolerance Databank, retailer own-brand lists and Coeliac UK (gluten-free). These lists can be very useful; however, consumers must be aware of the currency of the data.

Declaring the origin of ingredients is important to allergy sufferers. For example, lecithin comes from a variety of sources, including egg or soya, which will be vital information to consumers allergic to either of these foods.

Advisory labelling, i.e. "may contain" should be used only after a full HACCP audit.

It is also important to use the right packaging for the product. This may seem an obvious statement, but there have been several recalls of products caused by products being put in the wrong packaging. Old packaging should be discarded very quickly so that is it not used if the product formulation is changed.

Many allergic consumers establish which products are suitable for them and tend to use these products regularly; there are obviously implications in changing a recipe even though the change of ingredient is stated in the ingredients list. If a customer has been buying the product regularly, he or she might not realise that there has been a product change. To counter any possible confusion, an on-pack flash should be used stating "new recipe" or "new formulation" to alert allergic consumers to the fact that there may be a change to the allergen status of the product.

Labelling of promotional products is an area of concern to allergy sufferers. Promotional products are scaled-down trial-sized versions of products and therefore there is less space on the pack for ingredients labelling.

13.6   Recalls

Most big companies have recall policies. Advance planning for an allergen crisis is needed to ensure that a recall is conducted swiftly and effectively should a problem arise. There are several kits available for testing for the presence of allergens such as milk, egg, gluten, soya, peanut and hazelnut, should a problem arise with allergen cross-contamination. There are test kits for other allergens still in development, including a multiple allergen screen that has been developed at Leatherhead Food Research Association, and these kits can be quite useful for getting quick results for several allergens at a time.

13.6.1 Recall of product

The point at which a product should be recalled is controversial. It has been suggested in the US that companies should recall a product if there is more than 10 ppm of an allergen present. It could be argued that recalls should be instigated at lower levels of contamination however, because of the high level of sensitivity of some of the allergen testing kits now available.

13.6.2 Customer alerting

Customers should be alerted as quickly and widely as possible to a product recall using the media. The Anaphylaxis Campaign also posts alerts on the recall status of products from companies on its Web site.

13.7   Conclusions

This chapter has outlined some of the main issues faced by the food industry in relation to food allergens. It is important to be aware that peanuts and nuts are not the only issue in relation to food allergy. Many companies have improved their operating procedures and traceability in relation to tree nuts and peanuts, but the other allergens in the list of the top eight should not be ignored.

14. RETAILER PERSPECTIVE ON FOOD ALLERGY

Jenny Roberts

14.1   Introduction

This chapter will consider the retailer's perspective of food allergy and its implications, and will provide a brief overview of Iceland as a company and discuss the work that has been done by Iceland to improve allergy information for its customers, including labelling and availability of information.

When shopping for someone or with someone who has a food allergy, particularly to nuts, it can be difficult to find information regarding food allergens on food packaging. Iceland is committed to helping food allergy sufferers to have access to this vital information more easily.

14.2   Big Food Group

Iceland Foods plc is part of the Big Food Group, along with Booker Cash and Carry and Woodwards Food Service. There are approximately 750 Iceland stores, which are conveniently located in local high streets. Iceland employs approximately 22,000 people throughout its stores, depots and central office, and there are approximately 1,200 own-brand products within Iceland stores.

Iceland was very proud to be the first retailer to achieve many initiatives, such as offering a nationwide home delivery shopping service, which has now been complemented by telephone and Internet shopping services.

14.3   Food you can Trust Campaign

"Food you can trust" is an ongoing campaign at Iceland and is geared towards listening to its customers. This campaign was the driving force behind the removal of genetically modified ingredients from all own-brand products, and Iceland was the first food retailer to make this commitment.

In October 1999, artificial colours and flavours were also removed from own-brand products. Removing artificial colours was quite easy for some products - for example, taking the artificial colour sunset yellow or quinoline yellow out of a sponge and replacing it with the natural colour lutein. Removal of artificial colour from some products was more problematic, such as glazed cherries, which contain an artificial colour called erythrosine, which gives the cherries a bright red colour. The natural colour in cherries is anthocyanins and is a much deeper red than erythrosine. Changing from erythrosine to anthocyanins caused the colour of the product to change markedly; however, it was felt that this change in appearance in the product was acceptable, and the Hyperactive Children Support Group welcomed the removal of artificial colours, particularly those that are linked to hyperactivity and conditions such as asthma.

The technical aspects of removal of artificial flavours again depended on individual products, and a new range of own-brand crisps has been launched, which do not contain the flavour enhancer monosodium glutamate.

Iceland launched advertising campaigns to promote the removal of genetically modified ingredients. One campaign stated: "We banned genetically modified ingredients from our food on the 1st May this year. We did it because we refused to produce food which we would not be happy for our children to eat and we did it because we trust our customers' instincts more than those of the food industry."

Another campaign relating to the removal of artificial colours simply says: "Colour in the food because we do not."

In January 2001, Iceland Foods gave a free copy of the "Cool Food Big Book" to nearly 25,000 primary schools in the UK. The Cool Food Big Book is a reading book about healthy eating and how to achieve a healthy diet. It includes sections on special diets and food allergy. It has sections on nutrients and asks questions such as "What is a nutrient? and How do you make up a meal?" It has the Balance of Good Health in the centre pages. Iceland worked with the British Dietetic Association and the Food Standards Agency to update this model.

The book was written for the Iceland Year of Promise, following an Iceland-commissioned Mintel Report. The book was written in conjunction with the British Dietetic Association and advice was taken from literacy writers to incorporate it into the National Curriculum. The book was very successful and Iceland received a lot of positive feedback, not only from schools, but also from health professionals, and relevant industry bodies.

14.4 Food Allergy

Iceland has done a lot of work on food allergies. Customer queries have shown that there is increasing demand for allergy information on food packaging, and Iceland feels that, as a retailer, it has a responsibility to provide useful information to customers, and to make sure that the information is clear and honest.

14.4.1 Classification of allergens

Iceland classifies the following foods as allergens: gluten and wheat, cows' milk, egg, soya, fish, shellfish, sulfites and garlic. Garlic is included in the list not as an allergen but more with respect to food aversion. Customers continually request that the presence of garlic in food be clearly labelled. Peanuts and tree nuts are dealt with separately.

14.4.2 Product reformulation

When creating a new product or changing the formulation of an existing product, potentially allergenic ingredients are considered carefully with respect to whether they are essential to the quality and functionality of the product. For example, the formulation of a cottage pie was changed from using rice flour as a thickener, to using a wheat flour. Changing the thickener to wheat flour would mean that the product is no longer gluten-free, so this change has been investigated and, if possible, rice flour will continue to be used to ensure that it is gluten-free. If the use of an allergen in a recipe is necessary, the product is labelled in one of two ways. The first way is by means of a statement that is positioned below the ingredients declaration and it is simply an exclamation mark together with "contains gluten" or "contains milk", or whatever the allergen may be. The second means of labelling is an ingredients declaration where any number of allergens can be listed. Allergens can sometimes be hidden within compound ingredients and, therefore, full breakdown of all ingredients is necessary to ensure that the labelling is accurate. For example, curry powder often contains wheat flour, which is an allergen.

14.4.3 Allergen labelling

Legislation currently states that, if a compound ingredient such as curry powder is present at less than 25% of the whole product, it does not need to be broken down in the ingredients panel. Iceland currently works to a 5% level for the breakdown of compound ingredients and allergens will be declared irrespective of their percentage in the finished product. Iceland also has a policy to use the most consumer-friendly term for food ingredients. For example, caseins will be referred to as "milk protein".

14.4.4 Free-from lists

Iceland also creates "free from" listings available for the following allergens: gluten, milk, egg, soya. Lists are also available for vegetarian and vegan foods. It is also possible to create lists using a combination of these allergens for those with multiple allergies, for example, milk and egg.

Any recipe changes to a product are highlighted with a "new recipe" flash on packaging and, also, when free-from listings are sent out to customers, an update sheet on products recently reformulated is provided.

The "free from" lists can be posted direct to customers and can also be found on the Iceland Web site in the form of a product search facility. A copy is also kept in stores and customer care departments to answer any product enquiries.

14.5   Nuts

In February 2000, the Chairman of Iceland Foods received a letter from an 8-year-old boy called Damien who has a nut allergy. His mother wrote an accompanying letter saying how much she loved shopping at Iceland and how she appreciated the "food you can trust" initiatives. However, since her son had developed nut allergy she no longer shopped at her local Iceland store, as the necessary information was not available. The Anaphylaxis Campaign was contacted for advice and the necessary procedures were put in place to rectify this situation.

The first step was to review Iceland's technical policy for nuts. The types of statement made on packaging were also investigated, and the classification of nuts was reviewed. The following are classified as nuts by Iceland: almonds, brazil nuts, cashew nuts, chestnut, coconut, hazelnuts, macadamia nuts, peanuts, pecan nuts, pistachio nuts, sesame seeds and walnuts. The list does include coconut, and, although this is not viewed as a major allergen, there is still a question mark over its severity as an allergen; therefore, until there is sufficient scientific evidence against its allergenicity, it will remain on the list as a precaution.

A Code of Practice for the handling of nuts was written, based on the British Retail Consortium Guidelines for the handling of nuts, but also tailoring it to Iceland suppliers. The Code of Practice covers areas such as de-bagging of raw materials, storing the raw materials, scheduling of production runs, the use of re-work, and the use of frying oils, and also stresses the need for dedicated storage utensils and containers.

A nut audit was designed to enforce the Code of Practice, and was issued to all Iceland own-brand suppliers to be completed in February 2001. The findings of the audit indicate the appropriate nut statement to be used on packaging.

Any nut statement on packaging will be declared immediately below the ingredients declaration. The first statement, "Contains nuts", is self-explanatory for the use of products containing nuts as classified by Iceland policy. The second statement, "Made in a production area that uses nuts", is for use with products that do not contain any nuts but there are nuts present in the production area for use in other products. The third statement, "Made in a production area where no nuts are present", is prefixed with a tick to show that it is a positive statement. If the supplier does not use any nut ingredients in the production area, they sign a pro forma to be able to use that statement on packaging.

The second statement and the third statement use quite similar wording, so the words "uses nuts" in the second statement and "no nuts" in the third statement are printed in bold to make them stand out more for our customers.

Iceland produced the first edition of the Nut List in May 2001. This is a list of products made in a production area where no nuts are present, and is now updated and issued four times a year. A copy of the listing was sent to Damien in May, thanking him for starting the initiative.

Hazel Gowland, from the Anaphylaxis Campaign, visited Iceland in early 2001 and reviewed all of the procedures in place with regard to the Iceland nut policy, and made suggestions for future listings, which have been taken on board.

There is a lot of work ongoing on expanding the scope of the Nut Listing. For example, the inclusion of ready meals in the listing has just been achieved by working closely with one of our own-brand suppliers.

14.6   Allergen Testing

There is a need for more comprehensive testing methods for allergens to improve due diligence and service to customers. For example, a wider range of tests is required for nuts, in addition to the tests for peanut and hazelnut that are currently in place. Iceland will be implementing a testing scheme for any allergy statement that is made on pack; there will be a roll-out procedure of regularly testing products that have a statement such as "contains gluten" or such as a "nut" statement on packaging, controlled by Iceland's on-site laboratory.

New means of communicating information regarding allergens and food allergy to customers is being investigated - for example, shelf edge tickets, in-store information, or further use of the Internet. This will be in addition to any customer comments received regarding Iceland labelling or provision of special food lists.

The research and initiatives currently in place at Iceland will also be implemented within other sectors of the Big Food Group - Woodwards Food Service, and Booker Cash and Carry - thereby passing on the advantages gained by Iceland customers to the foodservice and wholesaler sectors.

14.7   Conclusions

Allergy information can be found on Iceland own-brand food packaging, highlighting the presence of any allergen, and Iceland's customer care department can answer any queries and provide regular up-dates of the "free-from" listing. The Iceland Web site: www.iceland.co.uk has an allergy section on it, including information, contact numbers for relevant organisations, a question and answer section and a product search facility, and the free from list can actually be downloaded from the Web site.

Staff in any Iceland store are able to help with any enquiries and hold a copy of the free-from lists at each store to help with customer enquiries.

15. THE IMPACT OF COOKING ON PROTEIN ANTIGENICITY

Professor Paul Davis, C.M. Smales, D.C. James

15.1   Introduction

This chapter considers the effects of thermal processing on proteins in terms of their structure, and possible changes in allergenicity.

As far back as 500,000 years ago, so far as we can tell, man rejected homophagia, which is the consumption of uncooked food. Cereals in particular are unpalatable and indigestible in the raw state. The act of applying heat to foods causes chemical reactions in foods that produce a wide array of compounds, which cause problems for allergists. Cooking is widely used in the home and food industry because it results in new or modified flavours of foods, and changes in texture through protein aggregation. It also causes thickening, tenderising and crisping. Cooking affects functional properties of food in terms of emulsification; it also causes texturising, flavour binding, and gelation. All these changes are absolutely crucial to make the experience of eating food positive. Cooking improves digestibility; for example, it breaks up grains, it tenderises, and prepares components for proteolysis. Cooking is also essential for safety, because heat kills microbes and inactivates toxins. It is crucial for preservation; if food could not be heated, there would be no canning industry. Heat can change the physical form of food by drying and granulating; it can also change its appearance by browning and opacity.

15.2   Effects of Thermal Processing on Allergenicity

Can thermal processing make food safer and less allergenic? It is widely believed that it can. There is a tendency to believe that, if a protein structure is destroyed by heating or other means of denaturation, it will be less allergenic. There are some cases where thermal processing can decrease allergenicity and, when a protein is unravelled, any sections that were brought together by the folding of the protein (its tertiary structure) to form a conformational epitope, are torn apart and no longer function as an antigen, thereby destroying allergenicity. Any epitopes formed by linear sequences of amino acids (primary structure) are not affected by heat.

Of course, thermal processing will create new epitopes as well as destroying others because, in the process of unravelling one protein structure and disrupting an epitope, a new epitope could be revealed.

In its natural state, protein is folded and stabilised by hydrogen bonds and electrostatic interactions. Most of the hydrophobic parts of the molecule face inwards and the structure is stabilised with hydrophilic amino acids on the outside, and hydrophobic ones on the inside. When a protein structure is broken up, the hydrophobic amino acids may be repositioned to be on the outside of the molecule. In this state, the protein can be carried in oil more easily than the native protein. The same hydrophobic effects can make protein more easily absorbed across the gut mucosa and into the circulation. Therefore, changes associated with cooking also include making the proteins more easily accessible to the immune system.

The science of food allergy originated in a classic paper by Prausnitz & Kustner, which was written in 1921 (1). It was the first rigorous and meaningful study on food allergy, described in this case as "super sensitivity", and Kustner, one of the co-authors, was the allergic patient. The paper contains one of the best descriptions of the symptoms of an anaphylactic kind of reaction to a food allergen, summarised as follows: "After half an hour, itching of the scalp, neck, lower abdomen, dry sensation in the throat. Soon afterwards, swelling and congestion of the conjunctivi. Severe congestion and secretion of the respiratory mucus membranes. Intense fits of sneezing. Irritating cough. Hoarseness merging into aphonia and marked inspiratory disponea. The skin of the entire body, especially the face, becomes highly hyperaemic, and all over the skin of the body there appear numerous very itching wheals, one to two centimetres large, which show a marked tendency to confluence. After about two hours, heavy salivation starts and is followed by vomiting, after which the symptoms very gradually fade away."

Kustner was allergic to fish and it is important to note that he was allergic only to cooked fish. Raw fish had no impact on him at all. He was so highly sensitive that even if he used fish glue and got it on his fingers he would have a reaction. One would imagine that, there would have been many such reports in the literature since 1921 of patients allergic to the cooked or the non-cooked versions of particular food. In reality, it is rather hard to find examples reported in the literature of people who are allergic to only the cooked version or only to the raw version.

The conformational structure of protein starts to break down at between 55 and 70 °C as secondary and tertiary structures are lost. Disulfide bonds holding the 3-D structure of the protein in place are broken between 70 and 80 °C, and, between 80 and 90 °C, new disulfide bonds form in new places and it is at this point that protein cannot readily refold into its native state. Between 90 and 100 °C, hydrophobic amino acids that were buried within the 3-dimensional structure of the original protein are exposed and failing to interact favourably with the solvent water, tend to cause aggregation. All these steps produce desirable properties in the food; they are all properties associated with food that has been cooked, but the process irreversibly destroys the structure of the protein. At between 100 and 125 °C, covalent modification occurs. At this point, a wide range of new compounds is created as sugars, for example, are broken down and bind irreversibly onto the protein.

It is clear that thermal processing can change the structure of epitopes found in raw protein, making them unrecognisable to the antibody that would have previously bound to them. X-ray crystallography and NMR can be used to study the structure of proteins folded in their native state but these techniques can tell us little about the structure of denatured protein. Hens' egg lysozyme, a well-known allergen, has a well defined structure with a region of α-helix and β-sheet. When the protein is heated, it goes through various changes, which have been mapped by Van Gunsteren *et al.* (2), using structural, dynamic and energetic properties of proteins to predict, through molecular dynamics, what is going on.

15.2.1 Degradation of the structure of hens' egg lysozyme by heat

With the loss of stabilising intra-molecular interactions the protein undergoes thermal motion and shakes itself apart. Further on in the denaturation process, the protein becomes quite open and unstructured. In an intermediate state, there is still a small section of α-helix present and the β-sheet is still in place, but most of the original structure is lost. Yet further on, the protein reaches a completely unfolded state, from which it is unlikely to return.

In addition to protein unfolding by thermal processing, covalent modification can also take place. Compounds such as sugars, fats and polyphenols are also present in foods and interact with proteins, either during storage or when heated.

15.2.2 Covalent modification of protein

On cooling, the protein might return (at least partially) to its original shape, but this is less likely if it had become involved in covalent modification brought about by reactions with other food ingredients at elevated temperature. Even without extensive unfolding, chemical modification can change shapes and create or destroy epitopes.

Figure 15.1 shows the Maillard reaction, which is responsible for browning, and begins with an aldehyde molecule, such as a sugar, interacting with a protein. This interaction results in a Schiff's base intermediate form, in which the protein has a sugar attached to it. Oxidation is also taking place, resulting in the formation of dicarbonyl compounds. In addition to this, the Schiff base undergoes Amadori re-arrangement, to produce the protein glycation product, which can be oxidised to produce dicarbonyls. The whole process produces dicarbonyls, cross-linked proteins, and dicarbonyl protein end products. Cross-linked proteins usually aggregate and the whole complex mixture is known as "advanced glycation end products" (AGEs).

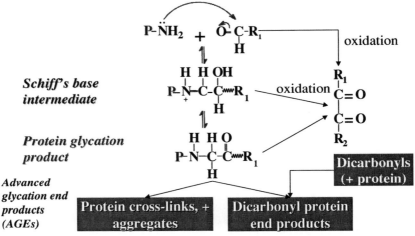

Fig. 15.1. Protein amino groups reacting with reducing sugars

This sort of chemistry occurs naturally in the body with ageing, and AGEs are produced spontaneously by non enzymic reaction.

Figure 15.2 shows a combination of the Maillard reaction and the effects of heating on proteins in food. First, sucrose is hydrolysed to free reducing sugars by heat, and native protein is unfolded. The free amino groups of unfolded protein can react with the reducing sugars, producing glycated protein intermediates (Schiff base adduct). Disulfide bonds are broken and reformed and free dicarbonyls are formed from the reducing sugars. Amadori re-arrangements result in glycated protein products, and random hydrophobic interactions between the unfolded proteins result in random dimers and aggregates. It has been shown that unfolded β-lactoglubulin can combine with unfolded α-lactalbumin to form a new structure, which is α-lactalbumin fused with β-lactoglubulin; the two denatured proteins can form a stable complex, which in some respects is different immunologically from its component parts (3).

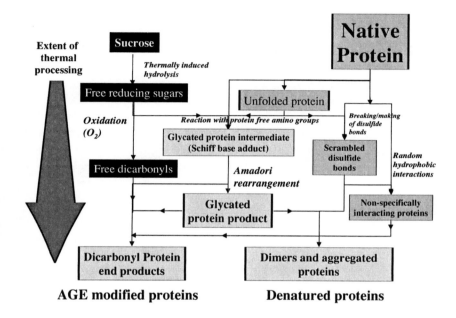

Fig. 15.2. Combination of Maillard reaction and effects of heating on sucrose in food

Glycated protein products can end up as dicarbonyl protein end products, dimers, and aggregated proteins, and the net result is a complex mixture of age-modified proteins and denatured proteins.

Cooked food contains a combination of the intermediate and end products of these reactions because some of the reactions are reversible and some do not go to completion. We do not know how much of each product is produced when food is cooked; however, the conditions that allowed one of these particular compounds to accumulate could be defined. When the food industry sets up a new process to make a new food, changes the recipe, or changes the cooking conditions, then it might well be that the result is a build-up of one or more of these compounds.

15.3   Effect of Cooking on Allergenicity of Peanut Proteins

This information is relevant to allergy. Maleki *et al.* published a study in 2000 in which they looked at the allergenicity of the peanut proteins Ara h1 and Ara h2 (4). The work was conducted using defined conditions to ensure that the proteins were modified in ways to ensure that the Maillard reaction was occurring, and AGE products were produced. The results of the study showed a 90-fold increase in peanut allergenicity after roasting. The roasted peanut proteins had an increased capacity to bind IgE antibodies from allergic patients and were markedly more resistant to digestion by gastric fluid, than were native proteins.

15.4   Effect of Heat on Protein Allergenicity

This is a topic of controversy because it is argued that the importance of the Maillard reaction in allergenicity (as distinct from just altered antigenicity) of proteins has not yet been proven. It is certainly difficult to find clear-cut clinical cases to prove or disprove the point. However, in immunochemical terms we can be precise about some of the antigenic changes. It can be proved that carboxymethyl lysine (CML) groups actually do attach to proteins when heated in the presence of sucrose. There is an area of science in which the thermal processing of therapeutic proteins such as those used in transfusion, for example factor VIII, is studied. It is important to preserve the integrity of these proteins; however, they require heat treatment to destroy viruses etc., and are normally heated in the presence of sugars. It has been found that, after heat treating specific proteins in the presence of sugar, the end result is a CML derivative that works as an epitope (4).

It is important to note, however, that if such CML-derived epitopes were allergenic, it would be common to find that patients were allergic to anything that was cooked, regardless of which specific proteins were present. The chemical structure modified in this reaction is an amino group with a positive charge, but the CML derivative that results is negatively charged, thus changing the whole physicochemical nature of that region. So a protein that has such a new epitope inserted may well become more antigenic in more subtle ways. For example, the insertion of CML through the cooking, and the presence of sugar may introduce new epitopes caused by the change in the charge of the protein and the altered molecular environment which this can bring about within particular modified regions.

What covalent modifications are important in allergenicity? Obviously, sugars are very conspicuous factors, but proteins can react with a wide range of other ingredients, such as polyphenols, present in many plant-derived foods.

Oxidised lipids may also be hitherto unrecognised factors in protein allergenicity. Fatty acids are highly flexible molecules, and it was originally thought that fatty acids could not be antigenic. However, if fatty acids are attached to proteins, they become epitopes and there are well characterised antibodies that specifically recognise defined fatty acids. It has been shown that fatty acids and products from heated fatty acids can attach themselves to proteins and cause an increase in allergenicity (5).

Oxidation through reactive oxygen intermediates can also expose hidden epitopes on proteins, as reported by Kalluri et al. (6).

15.5   Other Properties of Heat-Modified Proteins

As discussed above, a neo-antigen is not necessarily a neo-allergen because many other factors are involved in causing allergic responses.

Allergens have characteristics other than just antigenicity. An allergenic protein structure is typically more resistant to degradation, as found by Maleki (3). There are very strong indications that proteins that have had their hydrophobic amino acids re-oriented toward the outside, and that interact well with lipids, are likely to be absorbed more efficiently and might well enter the antigen-presenting cell more efficiently. A protein that is absorbed whole or as substantial fragments is more likely to be allergenic.

Majamaa & Isolauri (4) have supported this hypothesis in studies based on the model protein horseradish peroxidase (used because of its size and detectability). It was found that the protein tended to be absorbed intact across the mucosa of atopic children, but not absorbed by non-atopic children. Of course, this has more to do with the behaviour of the gut of atopic persons than the nature of horseradish peroxidase, but it clearly points to the importance of transmucosal antigen uptake in allergy. Different antigens are going to be more resistant, or less resistant to digestion, according to their structure and physical or chemical modification.

The next question is whether or not a modified structure can be recognised as a B-cell epitope? Not all modifications will be recognised; for example, CML is not likely to be recognised.

Does the absorbed molecule stimulate a TH2 response? If it does not stimulate a dominant TH2 response after absorption through the human gut it does not matter how antigenic it is in model systems. The interaction of a neoantigen with T-helper cells is, of course, a crucial determinant of allergy.

Is the modification reproducible enough to provoke and sustain allergic sensitisation? If the antigen is a minor component of the food consumed, and is not produced and consumed consistently, it is unlikely to be a significant allergen.

15.6 Clinical Case Studies of Allergy to Thermally Modified Protein

Is there evidence in the clinical literature that thermal processing and Maillard reactions are important? Some authors claim that these factors are important (7), but are not recognised very often. Malanin *et al.* reported anaphylaxis in an atopic girl after she had eaten cookies with pecan nuts present even though she was not allergic to raw pecan nuts (5). She had demonstrable IgE antibodies against epitopes in heated or aged pecan nuts, that were absent from fresh nuts. It was only when the nuts had been cooked or stored that they were allergenic. The neo-allergens appeared at around 2 weeks of storage or after roasting, strongly suggesting that a Maillard reaction product had caused the problem. Unfortunately, the published paper did not go on to characterise the exact nature of the epitopes.

Rosen *et al.* reported a patient who had no reaction to an extract of raw shrimp in a skin prick test, but an absolutely unequivocal, positive response to an extract of heated shrimp. This is another example of thermally dependent allergenicity (6), but of unknown chemistry.

Codina *et al.* reported finding an aero-allergen that was released from the hull of soya beans (7). The soya beans had been heated, either in storage or by fungal activity, or in processing, up to about 75 °C, and at least two new allergenic epitopes had been generated by the heating process. IgE antibodies were found to be present in allergic individuals not deliberately sensitised to heat-treated soya. This is another demonstration of the importance of neo-allergens in natural products.

Berrens reported an incident in which an extract from wheat flour was used as a hypo-sensitiser. A patient with bakers' yeast allergy was being hyposensitised with injections of wheat flour, which were fresh and colourless. In one of the desensitisation sessions, the patient was injected with an extract of wheat flour that had been wrongly stored for 6 months and was brown in colour. The patient suffered acute anaphylactic symptoms, which were not due to the wheat antigens as such, but to the Maillard-modified wheat proteins. It appeared that hydrolysed sugars had reacted with the wheat proteins during the storage period (8).

Bleumink and Berrens also reported a study in which β-lactoglobulin was heated at 50 °C with lactose and acquired a 100-fold increase in skin prick test activity (9). These sugar-modified proteins were subsequently found to interact with complement and had an enhanced IgE binding capacity (10). In 1912, it was noted by Kammann (11) that rye pollen acquired more skin test potency after supposed autolysis during storage.

This supports the view that AGE modifications create neo-antigens in situations other than in thermal processing, including food storage and transport, and this is an important factor to recognise. A lot can be learned from experiments with anti-viral heat treatment, modelling the so-called bio-processing of therapeutic proteins, such as in blood products. In these situations, the composition of the reactive mixtures of proteins and sugars are simpler and more clearly defined, making it easier to understand the results.

15.7 Conclusions

The effect of cooking on allergenicity of proteins is a highly complex subject. Patients, clinicians, ingredients suppliers, dietitians and, of course, food manufacturers should be made more aware of the need to consider not just the identity and source of ingredients but also the processing that they have received.

More informed and accurate risk analysis of new ingredients is required; novel proteins are being used in foods and new processes are being developed, so that both the pure native protein and the protein in the form in which it is eaten should be considered.

More effective approaches to minimising allergenic risk are possible through protein engineering, novel therapies and tolerogenic vaccines. There are enormously important lessons to be learned in this field, with the crucial point being to think about food as eaten rather than as pure proteins or unmodified ingredients.

Research in this area should provide a basis for more relevant, better controlled food extracts for use in diagnosis *in vivo* and *in vitro*. A great deal more research is needed on the question of AGE modification of proteins and allergenicity, and on the overall allergenic impact of cooking. This subject is complex, important and poorly understood. Almost every allergic reaction involving food as eaten in a cooked state is likely to involve neo-allergens without anyone being really aware. In most situations antibodies against proteins modified by heat and chemistry through cooking and storage are mixed up with antibodies against the native protein. Unless very detailed antigenic analysis is carried out it is almost impossible to be clear about what epitopes are involved in any allergic reaction to cooked proteins.

15.8 References

1. Prausnitz C., Kustner H. Studies on super sensitivity. *Centralbl.f.Bacteriol.1.Abt.Orig.*, 1921, 86, 160-9.

2. Van Gunsteren W.F., Hunenberger P.H., Kovacs H., Mark A.E., Schiffer C.A. Investigation of protein unfolding and stability by computer simulation. *Philos Trans R Soc Lond B Biol Sci.*, 1995, 29, 348, 1323, 49-59.

3.	Baer A., Oroz M., Blanc B. Serological studies on heat-induced interaction of α-lactalbumin and milk protein. *Journal of Dairy Research*, 1976, 43, 419-32.

4.	Smales C.M., Pepper D.S., James D.C. Mechanisms of protein modification during model antiviral heat-treatment bioprocessing of beta-lactoglobulin variant A in the presence of sucrose. *Biotechnol. Appl Biochem.*, 2000, 32, 109-19.

5.	Doke S., Nakamura R., Torii S. Allergenicity of food proteins interacted with oxidised lipids in soybean-sensitive individuals. *Agric. Biol. Chem.*, 1989, 53, 1231-5.

6.	Kalluri R., Cantley L.G., Kerjaschi D., Neilson E.G. Reactive Oxygen Species expose cryptic epitopes associated with Autoimmune Goodpasture Syndrome. *J. Biol. Chem.*, 2000, 275, 20027-32.

7.	Berrens L. Neoallergens in heated pecan nut: products of Maillard-type degradation? *Allergy*, 1996, 51, 277-8.

8.	Berrens L. Standardisation des allergènes par consommation du complément. *Ann. Med. Nancy*, 1977, Symposium IgE, 69-73.

9.	Bleumink E., Berrens L. Synthetic approaches to the biological activity of beta-lactoglobulin in human allergy to cows' milk. *Nature*, 1966, 29, 212, 61, 541-3.

10.	Berrens L., van Liempt P.M. Synthetic protein-sugar conjugates as models for the complement-inactivating property of atopic allergens. *Clinical and Experimental Immunology*, 1974, 17, 4, 703-7.

11.	Kamman O. Weitere Studien über das Pollentoxin. *Biochem Z*, 1912, 46, 151-69.

16. ARE NOVEL FOODS A SOURCE OF NOVEL ALLERGENS?

Dr Clare Mills

16.1   Introduction

The bulk of this chapter is based on the deliberations of a group brought together under an EU-funded concerted action project called PROTALL, which was coordinated from the Institute of Food Research in Norwich, and examined the relationship between the allergenic potential of plant food proteins, their molecular structures and their biological activities.

16.2   Definition of a Novel Food

The broad definition of a novel food is a food that has not hitherto been used for human consumption to a significant degree within the European Union (1). In real terms, if history were to be repeated, in the days of Sir Walter Raleigh, foods such as maize, potatoes, and chocolate would fall into this category.

The classification of novel foods is also applied to food ingredients, such as low-fat substitutes and nutraceuticals - substances added to food that might have a beneficial health effect. It also relates to novel food manufacturing processes - for example, very high pressure processing; i.e., for food that has been produced by high-pressure processing to be permitted for sale, it has to go through the novel foods regulatory procedure. The final group of foods classed as novel are genetically modified organisms (GMOs) for food use.

Novel foods have a history of being very successful - for example, quorn, which is based on fungal mycoprotein. Quorn is produced from the mould *Fusarium graminarium*, and a cause for concern in the early stages of development of this product was the fact that *Fusarium* moulds can produce mycotoxins. This is a product that has passed all the required safety tests and has actually made it all the way to the market place. There are no reported incidences of allergy to quorn to date.

16.3 Safety Testing of Novel Foods

16.3.1 Substantial equivalence

How do you tell if a novel food is safe? It could be compared with similar foods or ingredients that have a history of safe use. This might well be applicable to something like a new processing technique, where a product made by conventional means could be compared with a similar product that is made with a new process. Alternatively, the safety assessment can involve a toxicological evaluation of products such as fat substitutes, nutraceuticals, and compounds that might have a physiological function. Generally, it is clear which type of evaluation is appropriate, although there is much debate regarding the application of the assessment process to GMOs. One aspect of assessment of a novel food, particularly in the safety assessment of GMOs, is substantial equivalence. Substantial equivalence is the comparison of a novel food with a corresponding traditional foodstuff, should one exist. A traditional foodstuff is one that is already on the market and whose properties are as similar as possible to those of the novel food. There has been a lot of criticism recently about the concept of substantial equivalence, particularly from the Royal Society of Canada report (2). It is clear that substantial equivalence is a vague term that could be subject to ambiguity. New methods, such as genomics, proteomics, and metabolic profiling, will allow foods to be assessed in very great detail, and will make the concept of substantial equivalence more objective and more realistic in the future, as has been outlined by the UK Royal Society (3).

16.3.2 Allergenicity

Another part of the safety assessment of a novel food relates to its allergenicity. The problem with assessing the allergenicity of a novel food is that there are currently no good experimental models of allergenicity, which hampers the novel food assessment process. Similarly, it hampers endeavours to design therapeutics for food allergy because, without an animal model, there is no detailed knowledge of how food allergy actually arises at the molecular level. Food is made up of many components, and rarely is pure protein consumed. Chapter 15 describes the effects of the interaction of ingredients within foods, and the effects of cooking on the allergenicity of proteins.

Some years ago, the International Food Biotechnology Council, in conjunction with the International Life Sciences Institute (ILSI), produced a decision tree to aid this decision-making process (4). In the very early days of genetic modification, the first question asked during safety assessment of a novel protein was: "Is the gene an allergen?" If the answer was "yes", the next question would be: Does it actually react to human IgE? The decision tree followed various steps leading to a decision either that the allergenic gene could be put into a food so long as its presence was declared on the label, or that further consultation with a regulatory agency was required. This decision tree was established in 1996, and, not long after the model was launched, the biotechnology industry realised that it was unacceptable to put an allergenic gene into foods. The current situation is that, if the gene is an allergen, it is not used in GMOs for food use. Problems arise when a novel food is not obviously an allergen, and there must be an investigation into whether it might be an allergen. Factors such as the stability of the protein to digestion and processing are addressed, and animal models, such as the mouse, are used to assess whether the protein will stimulate the production of IgE. There are limitations to this process and, in 2001, the World Health Organization (WHO) established a working party that came up with a different decision tree (see Fig. 16.1). This decision tree heralds a more integrated approach to assessment of allergens.

The top of the WHO decision tree begins by asking the question: "Is the source of the gene an allergen?" rather than "Is the gene allergenic?" For example, does a new gene to be put into maize come from Brazil nuts, which we know are an allergen, or does it come from another source, such as a bacterium?

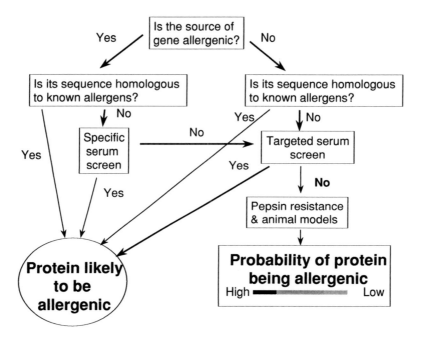

*Fig. 16.1. Assessment of the allergenic potential of foods derived
from biotechnology*
Source: FAO/WHO (5)

The next question relates to the sequence of the gene, and whether it is
homologous to an allergen. If the new gene is found to be homologous to a
known allergen, then the protein is likely to be allergenic, in which case it
cannot be used in food. If the gene is not homologous to a known allergen,
it is put through a specific serum screen using human allergic sera, and, if
this produces a positive result, the protein is likely to be allergenic and will
not be used in food. If the result is negative, the protein is put through
another serum screen from an animal model, and is tested for pepsin
resistance to give a probability of its allergenicity. The protein would only

be used in food if the probability that it was allergenic was low. There is a great deal of debate going on in the scientific community about the various different merits of this decision tree. The composition of the decision tree is open to adaptation, and it is likely that it will be modified as the science of food allergy evolves.

There are also many questions regarding the use of human serum in these kinds of testing screen. There are differences in the way in which the allergen testing process is viewed in Europe, compared with the United States. However, it is likely that some kind of consensus will be reached on allergen testing as a consequence of the debate that is currently ongoing.

16.4 PROTALL Work on the Decision Tree

The PROTALL group has been considering the WHO decision tree with a view to making it more effective.

To date, there is no history of novel foods, or foods produced using novel processes, having posed any greater risk of causing allergies than conventional foods. However, there is always room to improve the assurance that something new being introduced into the food arena that could cause an allergy would be detected. If there were better animal models, as there are for testing carcinogenicity, the testing process for allergenicity would be much easier.

16.4.1 Is the source of the novel gene allergenic?

The source of a new gene could be a food that may be consumed on a continent other than Europe, such as South America, and it is useful to find out whether the source of the new gene is an allergen in South America. One way to do this is to search the literature. However, a source of information such as Pubmed (http://www.ncbi.nlm.nih.gov/entrez/query.fcgi) contains, for example, over 5,000 references for wheat alone.

The PROTALL group has compiled an online plant protein allergen information database (http://www.ifrn.bbsrc.ac.uk/protall/). The database is searchable and contains information on the biochemical characteristics of allergens, together with clinical information. For example, there is quite a lot of information about rice and soya being allergens; however, it is surprising to note that the clinical history for some foods is very much poorer than might be anticipated - for example, there is not much clinical information available about rice allergy.

The PROTALL database is a tool to help in the evaluation of a new food, and a source of information about plant protein allergens. There are currently plans in place to expand the database to include all food allergens, and not just information on plant-based allergens.

16.4.2 Is the novel gene homologous to a known allergen?

There are many techniques to assess the sequence homology of a novel allergen compared with known allergens, and they use different approaches to establish how homologous they are. One of the ways of representing homology is in a tree formation not dissimilar to a family tree. For example, in looking at a particular class of proteins, such as non-specific lipid transfer proteins, the allergenic proteins can be classified, and the sequence homology between them established. The allergenic lipid transfer proteins have 70 to 80% homology. In assessing a new protein, its structure would be compared with that of proteins of known allergenicity, and, if levels of homology were high, it would be more likely to be allergenic.

16.5   Protein Families

A problem with relying on homology alone is that there are families of proteins that change their sequence, but not their structure and properties, including their allergenicity. This factor has been considered by the PROTALL group, i.e. whether plant proteins and plant protein allergens fall into particular groups and families.

16.5.1 The prolamin superfamily

The allergens in plants that cause food allergy by oral sensitisation generally belong to one of two so-called superfamilies. The first of these is called the prolamin superfamily, which has a characteristic signature in its sequence, which is a conserved cysteine skeleton (6). It is thought that the family has arisen from an ancestral gene during the evolution of flowering plants. This ancestral gene would have had the characteristic cysteine skeleton and, as time has gone by, plants have used the cysteine skeleton base to develop a number of different proteins, such as the lipid transfer protein; amylase/trypsin inhibitors; prolamin storage proteins (which are the storage proteins in cereals such as wheat, rye, barley and oats, which cause problems for coeliacs); or related proteins called 2S albumins, which also act as seed storage proteins. All of these proteins have the cysteine skeleton but they do not have identical amino acid sequences. The three-dimensional structures of many of these proteins have been determined, showing that they are conserved within proteins as diverse as the α-amylase inhibitor from wheat, 2S albumin napin from rapeseed, and the non-specific lipid transfer protein (Zea m 14) in maize. They all share a bundle of α-helices, held together by disulfide bonds formed by the cysteine skeleton, although the pattern of disulfide bonds differs. This is an example of conservation of structure without conservation of amino acid sequence. One of the problems with searching just for protein sequence homology is that anomalies such as this are not detected, and there is the potential for missing the identification of a potential allergen.

16.5.2 Cupins

The second allergen superfamily, called the cupins, again comprises allergens that sensitise via the gastrointestinal tract. The gene for this family originated in a prokaryotic ancestor, and the cupin family is so named because cupin is the Latin word for barrel, a characteristic of the cupin family being their barrel-like structure. This structural motif is present in proteins from fungi, called spherulins, which are found in spores, ferns also containing a vicilin-like protein (7). The cupin barrel has been duplicated in flowering plants, dycotyledenous plants and legumes; these are called bi-cupins, which are the vicilin-like globulins and the legumin-like globulins that are seed storage proteins. The vicilin-like globulin from peanut is called Ara h1 and is a major allergen, whilst glycinin is the 11S legumin-like allergen in soya. Despite the three-dimensional structures of proteins such as glycinin, Ara h3 (the 11S globulin of peanuts), coconut globulins, probably being almost identical, there is little similarity between their amino acid sequences.

16.6   Allergens in Cross-reactive Allergy Syndromes

Plant food allergens can be divided into those that sensitise via the gastrointestinal tract, such as the prolamin and cupin families, and those involved in cross-reactive allergies. The latter occur where someone suffers an allergic reaction to something - for example, pollen - and subsequently develops another related allergy to a fruit or a vegetable. Allergy to birch pollen is a good example of this as it is relatively common in Sweden, and 40% of people in Sweden with allergy to birch pollen also suffer from apple allergy. The symptoms of such allergies are very mild and include itching around the mouth, which is led to its being called oral allergy syndrome (OAS).

Bet v 1 is the major allergen in birch pollen and there are Bet v 1 homologues in carrot, apple, peach, and apricot. What is quite unusual is that there is a large surface area of the Bet v1 molecule that is homologous to Bet v 1-like proteins in other fruits and vegetables. Such homology on the surface of a protein indicates that Bet v1 must be very important in plant functioning; otherwise, it would not have such a conserved structure. Such levels of homology mean that, if someone develops an allergy to birch pollen, and produces IgE antibodies in response to this protein, there is a very high probability of their having IgE that recognises the same proteins in carrots or apples, etc. This is the molecular basis for cross-reactive allergies (8).

There are other cross-reactive allergy syndromes, a well-known one being latex. People who are exposed to latex on a routine basis - for example, health workers, who routinely wear latex gloves - may become sensitised to latex through inhalation of particles of latex from the insides of gloves that are coated with talcum powder. Cross-reactive allergens to latex are found in avocados and potatoes, because proteins homologous to latex allergens are found in these fruits and vegetables. Prohevein (Hev b 6.01) is the major allergen in latex. At the N-terminus, there is the hevein domain, which binds chitin, a polysaccharide found on the cell walls of fungi. The rubber tree may produce this protein to protect itself from fungal attack. Fruits such as avocado have an enzyme called a class 1 chitinase, which is produced when the avocado is attacked by fungi. Class 1 chitinases also have a chitin-binding domain, which is very similar to the hevein domain in latex (9). Therefore, if someone develops IgE antibodies to the hevein fragment in latex, there is a high possibility of that IgE recognising the same protein in vegetables such as avocados and potatoes, etc.

Therefore, it is important to identify proteins that are homologous to known allergens, in order to identify potential cross-reactive allergens; such birch pollen, and fruit, vegetable and latex allergens can be easily identified using this approach.

16.7   Pepsin Resistance

Pepsin resistance is a property considered in the WHO decision tree for allergen risk assessment (Fig. 16.1), and is determined by examining how quickly protein breaks down in response to pepsin, a protease that acts in the stomach and works at very low pH. This can be illustrated using a number of proteins. Using SDS PAGE, the allergenic Brazil nut 2S albumin, Ber e 1, has been shown to be fairly resistant to pepsin digestion, a property thought to predispose a protein to becoming an allergen.

There is some evidence that bovine serum albumin (BSA) is a milk allergen, albeit not a major one. BSA is far more easily digested than Ber e 1 under the same conditions. Haemoglobin was used as a control, because it is a protein that has not been found to be allergenic and was digested faster than both Ber e 1 and BSA, but was more resistant to digestion than previously expected.

The basis of this test is that, if a protein could resist digestion, it had more chance of being taken up by the gut and interacting with the immune system. However, we never eat pure proteins, and the transit time in the gut and in the stomach is several hours; the food matrix in which the protein is consumed will affect the rate of its digestion (10).

Marciani *et al.* (11) investigated the rate at which pepsin is mixed with stomach contents *in vivo* using magnetic resonance imaging, and found that, even in a liquid such as milk shake, it takes 2 or 3 hours for the pepsin to reach the middle of the milk shake to start potentially mixing it, and breaking it down, and not long after that period of time the stomach contents empty into the duodenum. Therefore, there are some issues as to the usefulness of data relating to pepsin digestibility of proteins.

SDS PAGE does not take any account of the fact that a lot of proteins exist as large molecules, and might fail to detect smaller existing subunits of proteins that have not been fully digested. For example, glycinin, the allergen from soya bean, is a large protein made up of several subunits. It can hold itself together very effectively, even after having been clipped in many different places (12). Disulfide bonds, such as those present in the proteins of the prolamin superfamily, are also very effective in holding the protein structure together following proteolysis.

Maillard browning products, and intermediary products of the effects of cooking on proteins may also affect the way in which proteins are digested by proteases. The effects of cooking may enable a protein to resist digestion much more effectively, even after it has been attacked by pepsin, and so to travel further down in the gut, be absorbed and interact with the immune system more effectively (13).

The food that we eat is a mixture of proteins and lipids, and forms an emulsion in the stomach. It is thought that proteins may also be able to evade being digested by adsorbing to the oil or to the emulsion interface in the stomach. It has been established that β-casein is digested very quickly by pepsin; however, if oil is added, digestion takes much longer, and quite large regions of the protein will stay stuck on the oil phase and evade digestion by proteases. It is not known how adsorption of protein onto lipid particles affects the mechanism of allergen absorption through the gut. The mechanism of lipid adsorption by the gut is also not known; lipids may well carry allergens across the intestinal lumen and may then interact more effectively with the immune system. It is not known how proteins in foods interact in the stomach and how that changes when the food passes from the gut into the duodenum.

16.8   Conclusions

Combined homology and protein family searching will become increasingly important in the future in order to improve the assessment of novel proteins for allergenicity, and ensure that novel allergens are not introduced at an early stage in the assessment process. Development of improved *in vitro* digestion models in combination with models of the mucosal immune system, and using animal models, will provide even greater assurance that novel allergens will not be present in novel foods than is presently the case.

16.9   Acknowledgements

Contributors to the project (Protall EU CT-98) were:

John Jenkins, Jim Robertson, Annette Fillery-Travis, Claudio Nicoletti, and Andrew Walker, Institute of Food Research, Norwich. The PROTALL management team: Peter Shewry from Long Ashton Research Station, Bristol; Charlotte Madsen, National Food Agency, Denmark, and Harry Wichers from ATO-DLO, Wageningen, the Netherlands. All of the other partners of which there are about 30 in all.

16.10   References

1.   Tomlinson N. The EC novel foods Regulation - a UK perspective. *Food Addit. Contam.*, 1998,15(1),1-9.

2.   Royal Society of Canada. Elements of Precaution: Recommendations of the regulation of food biotechnology in Canada. *Royal Society of Canada, Ottawa* 2001.

3.   Royal Society. Genetically modified plants for food use and human health – an update. *Royal Society London*, 2002, Policy Document 4/02.

4.   Metcalfe D.D., Astwood J.D., Townsend R., Sampson H.A., Taylor S.L., Fuchs R.L. Assessment of the allergenic potential of foods derived from genetically engineered crop plants. *Crit Rev Food Sci Nutr.*, 1996, 36 (Suppl), S165-186.

5.   Evaluation of Allergenicity of Genetically Modified Foods, FAO/WHO Expert Consultation on Foods Derived from Biotechnology, Rome 22-25 January 2001. FAO, Rome, Italy

6.   Shewry P.R., Jenkins J., Beaudoin F. Mills ENC. The Classification, Functions and Evolutionary Relationships of Plant Proteins in Relation to Food Allergies. *Biochemical Society Transactions*, 2002, in press

7.   Dunwell J.M. Cupins: a new superfamily of functionally diverse proteins that include germins and plant storage proteins. *Biotechnology and Genetic Engineering Reviews*, 1998, 15, 1-32.

8.   Ebner C., Hirschwehr R., Bauer L., Breiteneder H., Valenta R., Hoffmann K., Krebitz M., Kraft D., Scheiner O. Identification of allergens in apple, pear, celery, carrot and potato: cross-reactivity with pollen allergens. *Monographs in Allergy*, 1996, 32, 73-7.

9.   Breiteneder H., Ebner C. Molecular and biochemical classification of plant-derived food allergens. *Journal of Allergy and Clinical Immunology*, 2000, 106 (1 Pt 1), 27-36.

10.   Astwood J.D., Leach J.N., Fuchs R.L. Stability of food allergens to digestion in vitro. *Nat Biotechnol.*, 1996, 14(10), 1269-73.

11.   Marciani L., Gowland P.A., Spiller R.C., Manoj P., Moore R.J., Young P., Fillery-Travis A.J. Effect of meal viscosity and nutrients on satiety, intragastric dilution, and emptying assessed by MRI. *American Journal of Physiol Gastrointest Liver Physiol.*, 2001, 280, 6, 1227-33.

12.   Kamata Y., Shibasaki K. Formation of digestion intermediate of glycinin. *Agricultural Biological Chemistry*, 1978, 42, 12, 2323-29.

13.   Maleki S.J., Chung S.Y., Champagne E.T., Raufman J.P. The effects of roasting on the allergenic properties of peanut proteins. *J. Allergy Clin. Immunol.*, 2000, 106 (4), 763-8.

17. THE DETECTION OF TRACE AMOUNTS OF ALLERGENIC PROTEINS IN FOOD – TOWARDS THE DEVELOPMENT OF A DIPSTICK ASSAY

Dr Sabine Baumgartner

17.1 Introduction

This chapter explores the work currently underway at the Institute for Agrobiotechnology in Tulln, Austria, to develop quick tests for proteins or allergens.

The proteins for which tests are being developed include egg and peanut, or tree nuts - for example, hazelnuts. The allergenic proteins were extracted from the foods, and underwent further preparation for immunisation of chickens and rabbits; rabbits were immunised with egg proteins, and chickens were immunised with peanut and tree nut proteins.

Blood samples were taken from immunised rabbits, and eggs from the immunised chickens to obtain antibodies, which were used to develop our quick dipstick tests.

17.2 Development of the Assay for Egg Proteins

In order to prepare an in-house standard for the detection of egg in food, eggs were boiled and the egg white separated from the egg yolk. The egg yolk was then defatted with hexane three times. After drying, the yolk was mixed with the egg white and extracted with phosphate-buffered saline to produce whole egg protein extract, and, after centrifugation, was stored frozen in aliquots.

Antibodies are required to develop an immunological test, and the extracted egg proteins were used to immunise rabbits.

Three different egg proteins were used: whole egg protein, egg white protein and ovalbumin, which was bought from a local supplier. The proteins were mixed with Freund's adjuvants and injected subcutaneously into rabbits. Booster injections were given 4 weeks later, and titres were monitored.

IgG antibodies against egg used in this study were a gift from Dr Yupiter Yeung from the National Food Processors Agency in Washington, USA and were not prepared at the IFA-Tulln.

17.2.1 Preparation of samples for testing

Foods that were tested for the presence of egg included pasta, biscuits, puddings, dressings, and breaded frozen food. Ten-gram samples of these foods were mixed for 30 minutes with 100 ml of extraction buffer of phosphate-buffered saline at pH 7.4, with Tween 20 and Triton X-100. The solution was filtered and centrifuged, and the solution was ready to test for the presence of egg.

Sandwich enzyme-linked immunosorbent assay (ELISA) was the format used for the tests. Egg antibodies were coupled with horseradish peroxidase (HRP) by activating HRP with glutaraldehyde and, after a cleaning step, this was mixed with the egg antibodies.

17.2.2 Dipstick composition

A dipstick was developed used comprising a plastic strip with a nitrocellulose membrane attached using double adhesive tape. The dipstick is suitable for use in testing quite small amounts of solution in a test tube.

The sandwich ELISA consisted of two coatings of 1 µl antibody solution on the membrane, which was then blocked with skimmed milk powder. The dipstick was then incubated for 30 minutes with the protein extracts from the test foods, and for a further 30 minutes with the HRP-labelled antibody, followed by a colouring reaction with tetramethylbenzidine (TMB) for between 3 and 5 minutes. The total reaction time was one hour, and further work is required to reduce the time required to produce a result to the dipstick test.

An egg standard was used ranging from a low to a high concentration, and results were compared with those from the analysis of the food samples to enable an estimate of the concentration of egg present in the food samples. The problem with the dipstick test is that it is not possible to perform a so-called calibration for this test as would be done with a microtitre ELISA plate.

To verify the results of the dipstick testing, the food samples were also analysed using SDS PAGE and immunoblot, for example.

Garlic dressing that did not contain any egg was analysed by SDS PAGE and immunoblot. No egg protein was detected by either method, confirming that the results of the dipstick assay were accurate.

17.2.3 Limit of detection

In order to estimate the sensitivity of the dipstick, two different matrixes, butter cookies, and garlic dressing, were spiked with five different concentrations of egg protein. All the samples were extracted and the assays were repeated 10 times. The dipstick detected egg at 20 mg per kg, and at 2 mg per kg no egg protein could be detected.

Further testing is underway to obtain an exact estimation of the limit of detection of the dipstick assay.

17.3   Development of the Assay for Nut Proteins

In order to extract protein from hazelnuts and peanuts, samples were defatted with hexane, mixed with extraction buffer and phosphate-buffered saline, and centrifuged. Approximately 20 mg protein per ml were extracted. The extraction solution was analysed for protein concentration because chickens were immunised with it. Derco brown laying hens were injected with protein extract diluted to 1 mg per ml and mixed with Freund's adjuvant. Three immunisations were conducted at 6-week intervals.

Whereas, in rabbits, the antibodies to the egg protein were found in the blood, in chickens, IgY antibodies to nut proteins were obtained from the yolk of eggs laid by the chickens.

The chickens laid one egg per day, which were combined into batches of five eggs, whereupon the egg yolk was separated from the egg white. The egg yolk was then flushed with distilled water to remove all the egg white, and great care was taken because the membrane of the egg yolk is quite rigid. The membrane of the egg yolk was then punctured, and collected in a measuring cylinder.

The five egg yolks were diluted with phosphate-buffered saline in a ratio of 1:2. The fat content of the egg yolk was precipitated with 3.5% polyethylene glycol and centrifuged, and the supernatant was then frozen in aliquots.

In order to test that the chickens immunised with the peanut and hazelnut proteins had developed antibodies to these proteins, spot tests were conducted with a dipstick consisting again of a nitrocellulose membrane coated with the antigens: hazelnut and peanut proteins. The IgY antibody from the immunised chickens was used against the nut proteins, together with a commercially available rabbit anti-chicken IgG antibody that was already combined with HRP.

The spotting tests, using dilutions of antigens from 1:1000 and 1:50,000 showed that the hazelnut antibody was not as good at detecting the presence of antigens as the peanut antibody. It was also found that the peanut antibody cross-reacted with hazelnut protein.

There is much work to be done to develop these tests. The antibodies developed at the IFA-Tulln will be coupled with HRP, and the format for the test requires evaluation to determine the most sensitive type of testing method.

Work is planned in 2002 to develop dipstick tests for peanut and hazelnut proteins.

17.4 Acknowledgement

Thanks are extended to Dr Hermann, Dr Jupiter Yeung, and the Hochschuljubiläumsstiftung der Stadt Wien.

INDEX